The Science of God Volume 5

The Science of God
Volume 5
The Deluge
Boats, Floods, and Noah

R Lindemann

Aleph Publications
Wisconsin, USA

The Science of God Volume 5
Boats, Floods, and Noah - The Deluge
Copyright 2021 - R Lindemann ©
All Rights Reserved. Published 2023

Aleph Publications
Manitowoc WI

Paperback Edition
ISBN13: 978-1-956814-32-3

33 32 31 30 29 28 27 26 25 24   2 3 4 5 6

Disclaimer

All information, views, thoughts, and opinions expressed herein are those of the author(s) and are being presented only for your consideration and should not be interpreted as advice to take any action. Any action you take with regard to implementing or not implementing the information, views, thoughts, and opinions contained within this published work is your own responsibility. Under no circumstances are distributor(s) and/or publisher(s) and/or author(s) of this work liable for any of your actions.

Anyone, especially those who have been victim of misdirected explanation and understanding, may be best served seeking wise counsel before deciding to implement any information, views, thoughts, opinions, or anything else that is offered for your consideration in this work. All information, views, thoughts, and opinions in this work are not advice, directive, recommendation, counsel, or any other indication for anyone to take any action. All information, views, thoughts, and opinions offered herein are offered only as suggestions for your personal consideration, which is done of your own free will. Your life is your own responsibility; use it wisely.

Any use of trade names or mention of commercial sources is for informational purposes only and does not imply endorsement or affiliation.

Please note that most of the items in quotes in this book are from various versions of the Bible and may have been paraphrased.

Dedication

To start this topic off, we must consider that if the Bible is a True and accurate account of reality, then Noah is the father of us all. So this book must be dedicated to the obvious—Noah of Noah's Ark. But to add to this dedication, this book is also dedicated to all who are seeking the pure truths about the flood and of evolution and/or Creation. Keep your eyes on *Truth without agenda* and you will be amazed at what you might find!

Contents

Figures

Acknowledgements

Thanks to everyone who, over the decades, has contributed thoughts to the conversations about the Biblical flood. Your thoughts have been considered whether we spoke directly or I read your work. And thank you to all who are a part of the worldwide workforce that allows books such as this one to be created, printed, and distributed around the world.

Introduction

Most people don't realize that there is an association between the possibility of a worldwide Biblical deluge and the evolution-versus-Creation topic. The core of the matter really comes down to whether or not an actual global flood occurred as described in the Bible's Book of Genesis. Did the entire globe flood during the time of the Biblical Noah?

While this book is about the Biblical flood, it does touch on the debate regarding evolution-versus-Creation, and you will see why as you read on. Within the ranks of each side, there is a great deal of conflict on all sides of this flood topic. One side leans towards hostility while insisting that neither a global flood nor Creation is possible, while many on the other side will accept nothing less than six-twenty-four-hour-day Creation. Is there any compromise in between? And if so, then should there be a compromise?

Honestly, compromise is out of the question in this debate. Our quest is not to be agreeable for the sake of being agreeable; but rather, our quest should always be to understand the Truth about the details of ideas such as evolution, Creation, and especially a worldwide flood. The topics surrounding all of this range from the very beginning of existence and include big bang, evolution, astrophysics, physics, biology, geology, study of Genesis, floods, Creation of "kinds", and more. All of these topics surround the Biblical global flood—*if* it actually occurred at all. Each of the topics is a study in itself, which is why *The Science Of God* volumes are broken up into basic interest groups:

1.   *The Science Of God Volume 1 - The First Four Days*

The topics regarding the physical realities that are required for all things to be in existence are contained in the books just mentioned in order to help everyone test the validity of the scientific and Biblical claims.

The first four of *The Science Of God* books are a bit different than *The Science Of God Volume 5 - Boats, Floods, and Noah - The Deluge* because they do not concern the activities of man, with the exception of, finally, the arrival of Adam and Eve. *The Science Of God Volume 4 - Day Six - Evolution versus Man - In Our Image* is mostly only concerned about how Adam and Eve might have been created without violating common sense, logic, and everything we know about biology from observation.

The Flood of Noah's time is different because it includes elements of humanity, such as, who built the Ark-boat? How big was it? Could a boat that large endure rough seas? And so on. The flood is a much more nuanced topic, in that discussing the topic will lightly touch on big bang, geology, evolution, Creation, a bit of basic physics, mathematical realities, and various elements of humanity, but all in simple easy-to understand language. Since none of man existed until Adam arrived in the Bible's text, man has no impact on anything before Adam. However, after Adam's Creation, the Bible then ultimately becomes about man's journey from innocence to corruption to destruction to Salvation. But in this book, we are only concerned of about the flood facts and if the flood could legitimately have been a scientific reality, and if so, then how does that impact the idea of Darwinian-evolution?

Chapter 1

Impending Doom

What is the scientific possibility of a global flood? Could all of the mountains ever have been covered with water? Could a wooden boat—an "Ark"—handle rough seas? Could Noah have built the Ark all by himself?

There are so many more questions that can be asked about this topic that it is difficult to get them all in a single book of reasonable length. In the discourse within this book, there is no need to question if God exists regarding the flood. That will be assumed throughout this book since we are discussing the Biblical flood. Our goal here is not to test if there is a God, but rather since God does exist, is it scientifically plausible that a global catastrophe—the flood of Noah's time—wiped out humanity and covered the highest mountains with water to fifteen cubits above them?

To get any discussion about the Biblical flood of Noah's time underway, we first have to briefly set the stage of why this allegedly occurred. Debates surrounding the Biblical flood topic tend to get off track quickly during person-to-person dialog. So here, we'll make some assumptions since it is *the flood of the Bible* that we are analyzing. The primary assumption is that there is a discerning Creator-God who created all things and this Creator has very good reasons for doing things, such as flooding the entire globe. Without that assumption there is no need to bother discussing or even thinking about the topic. If there is no God, then there was obviously no Biblical flood.

What we are trying to understand is not if God exists, but rather, since God does exist, then is it reasonable or scientifically plausible to believe things such as that a worldwide flood could have occurred, and further, could Noah have built an Ark-boat with enough room to load every species on board the Ark? And, could the Ark have weathered the storm? As we proceed, you will see how all of this is also tightly connected to the evolution-versus-Creation debate.

The Bible claims that well over a thousand years after Adam and Eve where exiled from Paradise, the people of the world had corrupted themselves so terribly that God would no longer endure man's evil ways. God spoke to Noah in some manner and instructed him to build an Ark of "Gopher wood" because God was going to flush the filth from Earth with a worldwide flood. Some people have a problem with the idea that God would drown everyone just because they were behaving badly, but this was not without warning. It likely took Noah decades to build the Ark-boat that God told him to build, so the people knew what he was doing and why he was doing it—*That* was their warning. This means that they had decades to wake up and change their ways, yet they would not heed God's warning through Noah. To remove this point of discussion from the topic, we will also assume that God had a justified reason to create the flood in order to cleanse the Earth.

Some people argue that if man started from Adam and Eve, being only two people, then during the time before the flood they would not have been able to produce enough people to "populate the Earth". Thus, there would not have been many people and the entire world would not have needed to be flooded. That is an interesting perspective, but when you consider any major city and how quickly it increased in size in its early history, you then can see how quickly population can grow. Most cities started with a small group of settlers and within one- to two-hundred years the populations are in the millions. Now do that for about fifteen-hundred years and you will be looking at hundreds of millions of people. Then, add to that the fact, that according to the Bible, the people had lifespans of hundreds of years, thus allowing for incredibly fast population growth because there is very little die-off for the first several hundred years. We can safely assume that there was a substantial population that had corrupted itself, similar to what we see in modern times, but possibly even worse.

When a population grows quickly and a city falls away from God, corruption generally follows not long after. At least that's the experience of the nineteenth, twentieth, and twenty-first centuries around the world. The point to this is that the people had become so corrupted with all sorts of heinous activity to a point that God would no longer tolerate their terrible behavior. For that reason, God the Creator wanted to cleanse the Earth of the corruption. As it is understood, Noah and those from whom he descended all followed God with righteous fervor. And that righteous fervor filtered down from Adam, with whom God made an agreement, all the way down to Noah. Here is why God took this course of action:

Taken from Douay Rheims English Bible Genesis Six

"And after that men began to be multiplied upon the earth, and daughters were born to them, The sons of God seeing the daughters of men, that they were fair, took to themselves wives of all which they chose. And God said: My spirit shall not remain in man forever, because he is flesh, and his days shall be a hundred and twenty years. Now giants were upon the earth in those days because after the sons of God went in to the daughters of men, and they brought forth children, these are the mighty men of old, men of renown. And God seeing that the wickedness of men was great on the earth, and that all the thought of their heart was bent upon evil at all times. It repented him that he had made man on the earth. And being touched inwardly with sorrow of heart, He said: I will destroy man, whom I have created, from the face of the earth, from man even to beasts, from the creeping thing even to the fowls of the air, for it repents me that I have made them."

To avoid any dissenting thoughts, we are not only going to assume that God exists, but also that the people did, in fact, deserve to be destroyed in God's view, because their actions were that bad and were worthy of death.

Consider the Universe

When we accept the idea of a Creator who created all things, we must consider the Universe and how the Creator would have been pleased with all of Creation, as in the statements "And God saw that it was good." It is logical, then, that God would not tolerate people who corrupted themselves and constantly misbehaved and turned against the Creator of all things. And it stands to reason that any human who unjustly puts other created things at risk of destruction is a risk to parts of God's Creation. The logic here is that if you built a home for someone, for free, and then gave it to them to live in, but they then proceed to kill their own family and burn the house to the ground, would this be received well by you? I'm going to go out on a limb here and assume that you would not be at all pleased.

We need to try to grasp the level of frustration and anger that God might have had towards man for the atrocities that we were committing. The devastation would need to be complete to a point that there would be no possibility of allowing any of the evildoers to live through the flood or have any obvious evidence of that era left standing after the flood was completely done.

Because this book's primary focus is the plausibility of the flood events, we must assume at this point in the story that God had just cause and was very serious about this undertaking.

Advent of Pre-flood Civilization

When discussing civilizations and societies, the Bible is immediately at odds with evolution. The Bible alleges that man was made by God, in the Image of God, about six-thousand years ago, and that the first man was Adam from who came also, Eve, the first woman.

Evolution, on the other hand, alleges that over billions of years, microbiology somehow, by pure chance, got started, and from that, eventually all living things came to be by slowly and incrementally adaptively evolving through the adding of information to the creature form as the environment demanded in order for the survival of each particular lifeform. And then from generation to generation these small changes eventually caused man to exist in the form that we are today.

The difference between the Creation account and the evolution account could not be any wider. The Bible says man arrived whole and complete just as we are today, but evolution says that the life-forming amino acids could be as old as nine-billion years, but cells began to diverge only about two billion years ago, although there are many opinions about the time-frames in this regard. This is an area of great interest in the discussion. We have all of the geological strata that undeniably has entombed many creatures and recorded them in stone for us to study so many years later, and it appears that there is no foreseeable end to the fossil discoveries that we can make.

When it comes to civilization, we don't find much in the way of entombed fossils within the geological strata. Most of the activity of man is either still above ground or has, in some manner, become buried. But the depth of the found artifacts of man is relatively shallow, and the artifacts are typically buried in surface dirt rather than being entombed in rock. So, both the Bible and evolution agree that civilization is relatively recent. Evolution places the evolution of man beginning at about two-million years ago as a primitive creature that did not exhibit any major discernable level of sophistication or intellect, but then from that point on man became increasingly more intelligent.

the Bible, however, puts the arrival of man at only about six-thousand years ago being fully capable and discernable and intelligent from the get-go. In both cases, in relative terms to each belief set, civilization arose very quickly which is supported by the geological record as we understand it.

With our modern methods of tracking human DNA and migration patterns, we have been able to see what we believe is the migration trail left by the successive chain of our ancestors. Evolution places the so-called "cradle of civilization" in Africa, but the Bible places Noah's landing in a several hundred mile radius area surrounding the land of Israel.

One point that both sides of the debate agree on is that by six-thousand years ago man was fully developed to the levels that we are today. The advent of civilization is one key area that could make or break either side's case.

Prophecies and Predications

There are many prophecies in the Bible, but even evolution has its own predictions. Within their respective fields, prophecy and prediction are equivalent.

When it comes to prophecy in the Bible, there are two basic types to look for. One is that God or an angel came or appeared to someone and then told that person to tell the people a particular message as a warning for the people to whom the message was to be delivered. The other type of prophecy is a bit different because it was instruction to take certain actions. While there was some foretelling of events between God and Noah, the primary task for Noah was to believe God about the impending doom of a global flood that would eventually be coming, and then to follow the instructions to build the Ark. They were simple and without change, very as-a-matter-of-fact statements from God.

Science differs greatly from any type of prophecy in the Bible. Evolution-science is based upon people's imaginations from which they form speculations or assumptions and then go on to "prove" those assumptions by doing research, which is done mostly through archeology in regard to evolution.

One point of hidden contention between sides in the evolution-versus-Creation debate is the unfair tactic of applying your own constraints to the other side's theories. Both sides of the debate tend to do this where someone from the Creation side will say something to the order of "that can't be, because man was only made about six-thousand years ago." Or the evolutionist might say that the Earth has to be older the only six thousand years because the various layers of the geological strata are each millions of years old. Using such tactics might, in the long run, be complimentary to the points made by one or the other side, but you cannot constrain someone else's theories to the constraints of you own theory if you want to do an honest job of testing either theory.

When speaking of evolution, we must use the evolution model. And when speaking of Creation, we must use the Bible's Creation model. The prophecies and predictions can generally only work within each side's theories. At this point, it is important to note that both sides of the evolution-versus-Creation debate have more than one perspective to consider. For instance, there are many Creationists who believe that God created everything in six twenty-four-hour days. However, there are also those who believe that God did it, but that the days in the Bible were very long periods (See *The Science Of God Volume 1 - The First Four Days.)* There are similar divisions on the evolution side, but the timespans are considerably greater than on the Biblical side.

Perhaps the most prominent tactic in discussing this entire topic is to attempt to discredit the opponent's theory through incorrect association of one piece of their theory to other non-related items in the general beliefs of other people who also promote the theory. A theory's points have nothing to do with people's actions or thoughts or opinions—a theory is laid out on paper and those are the only points that matter in the debate regarding the theory. When we revert to constant ad-hominem attacks, can we ever hope to see the advance of accurate science? Not likely.

Trying to discredit the Bible by implying that is it "only stories" or "mere poetry" does not answer to the accuracy or inaccuracy of the events stated in the Bible. Or trying to discredit the Bible by saying that "The prophecies weren't actually proclaimed before the prophesied event since they are written long ago, and we were not there. Thus, we can't know for sure whether or not the prophecy was written after the fact, which would then make it to not be prophecy." It is in this same way that we cannot ever know for sure that *any* relatively current events occurred as someone may have related it to us, especially when there are two conflicting accounts. We have to be discerning and unbiased in our evaluation of **all** available data.

Since the big bang has various theories where the Universe is believed by some people to be expanding, and by other people it is believed to be contracting, and yet others believe it to be static, then by using the same logic as is often used regarding God or the Bible's validity, we would say that the big bang does not exist and could never have occurred. This anti-Bible logic seems to be so deeply rooted in the minds of the promoters of evolution logic, that we must also hold those evolution beliefs to the same scrutiny as the Bible supporters' beliefs are held to by the evolution supporters.

An Eye for an Eye

The tit-for-tat-gotcha game that is often played between sides only provokes conflict, rather than sound discourse. It is *an eye for an eye* mentality and it avails us little in finding the truth.

Before we get into the good stuff, we have to first remember that when discussing the details of the Genesis flood account, we must not use any of the rest of the Bible as evidence for the global flood, or evidence of an intelligibly guided Creation for that matter. Any Biblical text after those events actually occurred had no bearing on the events themselves. For instance, it's okay to reference Christ when quoting something like the "an eye for an eye" text, so long as the theory does not at all rest upon such quotes. Such as in Matthew Chapter 5 when Christ said "You have heard that it hath been said, An eye for an eye, and a tooth for a tooth. But I say to you not to resist evil: but if one strike thee on thy right cheek, turn to him also the other. And if a man will contend with thee in judgment, and take away thy coat, let go thy cloak also unto him."

Jesus took a greatly different approach than God took in Genesis Six. It seems that the text of "an eye for and eye" found in Exodus was indication that the punishment should fit the crime. The core of that lesson is, if you had damaged someone's eye, then they should not kill you, but rather only the equivalent in damage should occur and no more. When discussing the Bible, these sorts of issues often arise, and it's usually when one or the other side has no plausible argument with which to combat the opposition. These distractions stop us from getting to the root of the question: Could the flood have occurred as implied in the Bible? While discussing "an eye for an eye" does not affect the possibility of a worldwide flood, it does address the level of anger that God had towards the people to justify total destruction through a cataclysmic global flood.

It's pretty clear that if we stick to the topics at hand and stay clear of questioning God's existence, or the justification for total global destruction, we can then more concisely address the issues required to possibly arrive at reasonable conclusions regarding the scientific plausibility of the Biblical global flood. It is imperative that you understand, in this flood discussion, that the flood is a very big problem for evolution–and evolution is a very big problem for the flood. As currently understood by pop-science or by Bible believers, those two situations cannot both occur, therefore only one side can be correct. And quite honestly, if a global flood did occur, then the probability that we evolved is extremely low, especially when considering that evolution adherents proclaim long-age evolution to be "fact". This will all make more sense as you read on about the critical, but simple, scientific details that would be required for a global flood to occur and be capable of covering the highest mountains.

Chapter 2

The Animals

Another assumption we will make in this book is that an Ark was built by Noah. With that assumption, we can then investigate the feasibility of Noah building it to the size stated in the Bible. We can also investigate if all of the various "kinds" of animals could realistically fit in the Ark.

The animals are said to have been loaded on the Ark when two of each kind came to Noah at the appropriate time. Here is the text: "They and every beast according to its kind, and all the cattle in their kind, and every thing that moves upon the earth according to its kind, and every fowl according to its kind, all birds, and all that fly, Went in to Noah into the ark, two and two of all flesh, wherein was the breath of life." Here is where we get to one of the issues that come between a global flood, versus all creatures having evolved, and so here begins a major point of contention between debate sides. Evolution proponents argue that there are millions of "species" with estimates as high as one-hundred million species, but more reasonably it is believed to be fewer than ten million species. How, then, could all of those species possibly have fit on the Ark?

At one-hundred million it seems highly unlikely, but with only ten-million the chances are better, but not much better because ten-million is still a very large number of species, especially when you consider that two of every kind, twenty million creatures total, would have to safely fit into the Ark. This means then that the Ark would have to be huge! So, just how big was this boat?

In Genesis Six it says "The length of the ark shall be three hundred cubits: the breadth of it fifty cubits, and the height of it thirty cubits." That's 300 x 50 x 30 = 450,000 cubic cubits. According to the flood text in the Bible, the Ark had three levels, "Thou shalt make a window in the ark, and in a cubit shalt thou finish the top of it: and the door of the ark thou shalt set in the side: with lower, middle chambers and third stories shalt thou make it." This would offer 45,000 square cubits of floor space (300 length x 50 width x 3 floors = 45,000 square cubits). The problem that we have here is twofold. First, we are not sure of how long a "cubit" was at that time, and whether or not a cubit is more than a foot in length. How could one man possibly build such a huge boat? More on that later.

Estimates on a cubit length run somewhere between 12 inches and 21 inches as measured in our modern world, but it could be more or less than those estimates. This means that the Ark would have had square footage ranging between 45,000 square feet to 137,812 square feet, with total cubic footage ranging between 450,000 cubic feet to 2,411,718 cubic feet. The Ark was a massive structure by any standard using either cubit measure.

Could all of the various kinds of animals fit in the ark? To begin to test this, let's first take the lowest number of cubic feet and an estimate of species at ten-million. Since Noah took two of each kind on the Ark, that means we have to double the ten-million species to twenty-million total creatures. Let's divide 20,000,000 / 450,000 = 44. That means that Noah would have to pack 44 creatures into each cubic foot of a 300-foot-long Ark (About 91 meters long). Is this possible? Did he use magic? Let's dig a bit deeper into Noah's space logistics problems.

Classifications and Species

If you read *The Science Of God Volume 3 - Day Five and Day Six - The Creatures - Revolution or Evolution* you will be well familiar with the species problem that we face in science, especially with regard to the Bible's "kinds". Classifying species has been a contentious point even within the heart and mind of a single person doing the classifying. At some point, someone has to define the parameters of a specific "species", and while those parameters might be specifically stated somewhere, in the end it is a judgement call by the particular person who is qualifying any one species at hand. Two different species as defined by one person could be defined as a single species by another person, thus the ten-million species might be considerably more or less depending upon who is defining each species. In addition to this, the species number of ten-million is only an estimate. No one has actually counted them. The number of species is a very rough estimated figure and is not specifically known, nor is the estimate particularly scientific, rather it is simply an "educated" **guess**.

Then we have the issue of "species" versus Biblical "kinds". The modern English Bible's say nothing about "species" going into the Ark, so in that sense the distinction of species no longer has any relevance. Some people on the evolution side will even press this issue by demanding a definition for a Biblical "kind". However, there is this little-known secret; in the ancient Latin Bibles it uses the actual word "species" or the word "genus". This means that as far as Bible translation is concerned, "kind" and "species" have the same exact meaning and value. This places us back in the scientific realm of ten-million "species" or "kinds" with two of each.

One critical point to keep in mind is that Noah was to bring "every beast according to its kind, and all the cattle in their kind, and every thing that moves upon the earth according to its kind, and every fowl according to its kind, all birds, and all that fly" on to the Ark. This means that all species that are water-only creatures can be removed from the numbers, which is estimated to be about twenty-five percent of the total estimated species quantity. This immediately removes two and a half million species, causing Noah to only have to stuff 33 creatures in each cubic foot of space. There is a great deal more to this space problem that Noah was dealing with that we will get to later.

Darwin's Objectives

In Darwin's book *The Descent of Man*, he talks about his views on evolution and natural selection, which are two different things. It is through natural selection that things are believed to have evolved. But Darwin went so far as to say that he wanted to "overthrow the dogma of separate creations", which means that all creatures came from a single point source of evolution. This, for Darwin, is a stumbling block. He at one time was training to become a minister and he then believed the Bible's Creation account as *he* understood it. However, when he went abroad and witnessed very similar birds and other creatures, but with varying features, he came to realize the obvious—that animals can change over time. This striking revelation should not have been a striking revelation since people had been crossbreeding dogs for thousands of years to create different breeds that are unquestionably unique from their dog ancestors, and Darwin was well aware of that.

Once Darwin came upon his new-found revelation that animals changed through generations, he eventually completely abandoned his Creation beliefs and began to disregard them as nothing more than tall-tales invented by man. Were Darwin's objectives right, that there were no separate Creations of animals as stated in the Bible?

What was Really Going On

So as not to spend too much time on Darwin's evolution, we can sum this up by realizing that the definition of species is not "science", it is an art form that allows artistic license. This means that the species distinctions might be far too tightly defined since a wren with a longer beak and a wren with a shorter beak might be defined as two different species by someone, yet they are both "wrens". This point is particularly important, as is the speed at which animals change when they are crossbred. Darwin battled wits with people who opposed his sacrilegious thoughts regarding the evolution-versus-Creation debate, and in so doing, he dug in his heels in order to hold his ground, and it appears that he disposed of his Biblical objectivity.

Rather than realizing he might be wrong regarding what he specifically believed about Biblical Creation, he chose to use the same sort of rationale for his new-found natural selection evolution model of origins. His rationale has now become a part of international scientific beliefs, regardless of the efficacy of that rationale.

For some unknown reason, people on both sides tend to polarize on these issues, and because of that polarity they stop looking and listening. Darwin and most of his followers completely disregard the Bible, while overlooking the very obviously good advice in it. Life is not always the same as we perceive it; for instance, in the Bible, people were allegedly sacrificing their children, and for some unknown reason it needed to be explained to people that this was not good and is not what was intended when humans were created. Also overlooked in the Bible is that the Bible had a great amount of scientific foresight regarding disease and health. We often misunderstand the Bible because it also dealt with their health problems at hand during those times, many of which we generally no longer experience in our modern era because of the Bible.

The existence of the species is certain, however the classification of them is *not* certain. And it is the classification of species that is a contributing factor as to whether or not two of each kind could fit on the Ark. Darwin has either inadvertently or deliberately tainted the minds of many people with his objectives.

Evolutionism has come to mean that there is no God. And the species issue is one method evolution proponents use to make that point. If the flood account can be legitimately dismantled then so goes the rest of the Bible, and God along with it. Not all evolutionists are atheists, but it is a fairly strong trait among evolution proponents.

Chapter 3

Rising Tide

We will revisit the association of a global Biblical flood to the alleged evolution of animals in a later chapter, so let's pause those thoughts for a bit and consider the actual flood itself. Is it realistic to believe that the entire globe flooded as it is written in the Bible's Genesis where "The water was fifteen cubits higher than the mountains which it covered"? Is this possible? Could that much rain have fallen?

Calculations of the cubic miles of snow and ice in glaciers and on the polar ice caps estimate that if all of the ice and snow melted it would then raise all of the oceans by roughly about two-hundred-thirty feet by some accounts. This could not possibly cover the mountains that are miles high, only the base of some mountains would be covered at that depth, leaving miles yet uncovered. So here supporters of the Biblical flood are immediately faced with a hefty lack of water to do the mathematical task of covering the mountains to "fifteen cubits higher than the mountains which it covered"; and that is a serious problem for Biblical literalists.

Flood Odds

This book won't trouble you by listing all of the global flood stories from cultures all around the world, but know that there are many. In our modern era, we have the luxury of looking these things up in books or even electronically online to read the various global flood accounts. If you are interested, then search for ancient flood stories like the Bible tells of.

The odds that nearly all key cultures, both past and present, have tales of a global flood is a bit peculiar. If the flood account is invented, then for so many different cultures to tell a similar account of a worldwide flood with just a few survivors is, practically speaking, impossible odds—unless all cultures share a common early history. Having the key elements of the scores of existing flood stories from around the world all match so closely and then realizing that if the Biblical flood account is true, then it was only Noah and family who remained to repopulate the Earth, which is the unspoken claim of the Bible. Therefore, based upon those data bits, it is a reasonable and logical assumption to link the Biblical flood to all of the other similar flood stories. But is the Biblical flood of Noah's time the source of the other flood stories, or are some of the other flood stories where the Noah story comes from?

To answer this, we are limited to using logic and the cumulative flood stories. If you have read *Understanding the Bible - the Bible How-To Manual* AND *The Things We Don't See* and *Understanding The Church - Upon This Rock I Will Build My Church* you will already have an idea as to where a part of this is going. The Bible is not just some random "invented book of tales" as many would have you to believe. The Bible is recorded history much the way we record history today when we *write it down*—and that is what is lacking in the other flood stories. Many, if not all, of those other stories worked like the "telephone game" where people sit in a circle and whisper something in the ear of the person next to them, and as it is passed from person to person it changes. This occurs until it is finally delivered to the original person who first whispered the story, where the story often comes back with many elements altered or altogether left out. Without actually *writing down an account of an event*, that event's details will be slightly changed when going from person to person, and over enough generations of sharing such stories, the original story will be noticeably altered, yet it will typically still contain the most noteworthy parts, such as a worldwide flood, a boat, a few survivors, and animals, etc.

When you read the Bible's flood account, it is the most articulate and real sounding account of them all, versus the more cartoonish word-of-mouth accounts from around the world. Yet they all share the critical points of the Biblical flood. And since all people after the flood would have descended from Noah, all people would have cultures that would have been privy to that information, thus all cultures have a flood story.

The Bible is the written history of Original Sin and our Salvation from it, and it includes the lineage of the people through whom that Salvation was delivered. In the Bible, people were told to "write this down". The custom of the Israelites writing things down was handed to them by God, and to this day the Israelites are very articulate about doing so. Writing it down is why the Bible's flood account is fairly articulate and the other flood accounts are not. Writing history down preserves it with accuracy according to the facts at the time it is written, thus sealing the words for later generations to read and contemplate. In our modern world of electronic communications, our culture is at serious risk of entering and succumbing to the telephone game's flaws. When electronic information changes due to someone making so-called "corrections", we then lose the ability to know that the information was stable, where on the other hand, with written or printed information, it cannot be changed on the already printed paper.

Regardless of the flood being fully global or not, if we assume that Noah and family did survive some sort of cataclysmic flood that wiped out everyone else, that then explains the likely origin of all flood stories. Yet, we still have a dilemma regarding the serious lack of water, which according to modern era topography could not possibly cover all of the land, even **without** the mountains being considered at all. That is pretty damming for the belief that there was ever a global flood, and that is even if all of the flood stories matched to perfection. This is true, except for one indicative detail: Everywhere you dig deep on Earth appears to have had some sort of cataclysmic flood at some point in history. However, the fact that the entire globe shows evidence of flooding does not prove that a *global* flood occurred. Horrific regional floods occur on a regular basis, and over millions or even just thousands of years, most places have been flooded to some extent. This means that, theoretically, every point on Earth could have had a flood, but not necessarily all at the same time.

However, the mountains contradict regional flooding because there are fish fossils found high in the mountains. This presents a problem when discounting the flood for those who deny that a global flood could ever have occurred. Because the mountains are much higher than the median land altitude, the water would have at minimum had to have been at least as high as the location of the fish fossils found on the mountains when the flood that may have placed them there actually occurred. This possibly means that only the tops of the highest mountain could have been above flood waters in that case.

The fact that fish fossils are found high up on mountains helps to support a global flood, but it does not prove that all of the highest mountains where covered by water to at least "fifteen cubits" above every peak. So, the fish fossils do lend to a global flood based upon the fact that they are found well above current median land elevations.

This is where things get interesting and the Biblical flood gets a smack-down from science. Many of those who deny that a global flood could have occurred, rightfully point out that the mountains were not always there and that the fish fossils would have been formed before the mountains formed, because at some point in the past, those mountains where at the bottom of the seas. And this is indeed what the evidence shows. So, it seems that from this assessment, the global Biblical flood has been defeated since the fish fossil altitude has been taken out of the picture due to the fact that the mountains formed millions of years after the fossils formed. This is made clearly evident by the projected angles of the rock that the mountains are made of. Something caused a massive shift forcing the rock to buckle and become the mountains that we all know and love today.

A Powerful Planet

Now that we have shown that the fish fossils were likely not from a global flood that covered the mountains, we can rest easy knowing that a global flood is a bit of a reach... that is until you stop to realize that the fish fossil forms that are found in rock were at some point considerably lower in height. This means that at some point water did, in fact, cover those mountains before they were mountains; which means that the mountains at some point would have had to have been lower than the general land height. If the mountains were lower than the surrounding topography at any point in time, then if that surrounding topography was covered so was the material that the mountains are made of before the mountains were actually formed. Now at this point, those who disagree with the Biblical flood and deny that it could have ever occurred have just given the Biblical flood supporters perhaps the single most important piece of evidence possible, which is that they have proven through obvious and simple logic that the mountains as we see them today did not exist before they formed.

Such massive upheaval was catastrophic! Those who deny that a global flood occurred assert that the mountains took millions of years to form, which they claim happened due to the movement of the sea floor towards the land, thus forcing the land to be horizontally compressed causing the crust of the Earth to buckle, thus forcing the crust upwards forming the various mountain ranges. It sounds like a plausible theory that the land buckled from movement of the sea floor over millions of years, except for one looming point that tends to get ignored. If you have ever paid any attention to rock layers when driving through highway cut-ins that are cut through rock, you might have noticed that over your lifetime you will have witnessed some erosion and crumbling from freeze and thaw cycles. Yet when we look at the mountains, we see fairly sharp edges of the fractured rock. Eyewitness examination of the mountains and basic logic make it very apparent that the mountains buckled and were then forced up from extreme lateral pressures. The problem with the mountains being millions of years old is that they show erosion and weathering of only thousands of years. It appears that we might be in a sort of timing conflict between global flood enthusiasts, versus global flood deniers.

Fifteen Cubits over the Highest Mountain

As just discussed, there is no need to cover the mountains at their current height, because the claim can be made that the mountains were not there during the flood. However, here the flood deniers will assert that the current-day mountains formed millions of years ago, yet the flood is claimed to have occurred only around four-thousand-five-hundred years ago, therefore, the flood would have to have covered the mountains as we see them today. This rationale befuddles some flood enthusiasts because they buy into the millions-of-years-mountain-formation theory. But if you recall in an earlier chapter, it was mentioned that we cannot use the one side's constraints with the opposing side's theories to discredit them.

Flood enthusiasts need to realize that the mountains were not always there, which is made clear by simply observing them up close. However, the flood enthusiasts' befuddlement is due to timing issues. It is not provable that the mountains took millions of years to form. We simply do not know; thus, the flood supporters must assert that the mountains did not exist and were not a factor in the depth of the flood that would have required that the water be over the highest of the current-day mountains.

We also have to consider the language used in the Bible and the translation words used, as well as our perception of those words. The word "mountain" invokes in us a sense of extreme height given the view of mountain ridges that we see around the world today. But in the Bible, it says "And the waters prevailed beyond measure upon the earth: and all the high mountains under the whole heaven were covered. The water was fifteen cubits higher than the mountains which it covered." Yet there is no mention of the exact height of the mountains, rather it only indicates that they were covered by fifteen cubits of water. Nor do we know the size of those "mountains" regarding their surface area and volume. If the Bible used the term "mount" it would give us a very different picture. The term used in the Latin Bible was "montes" which does not indicate any particular size, but rather only a noticeably raised area. This is a very important point to understand in trying to get to the bottom of the details in this Biblical flood debate.

Chapter 4

The Water Will Find Its Way

If adhering to the Biblical flood perspective, and while viewing the actual mountainous evidence that stands before us today, we know that the materials that the mountains are made of were most certainly under water at one time due to the evidence of fish fossils that have been found on them. We also know that, logically speaking, the mountains were not always there. For our purposes here, we are going to ignore the speculative scientific theory that the mountains began to form *millions* of years ago and did so over the millions of years since then. Instead, here we will assume that they formed sometime during and/or after the Biblical flood which occurred within the last five-thousand years.

Ignoring the asserted modern-scientific time estimates to form the mountains, by assuming the pre-flood mountains were much lower, we can easily explain the fish on the mountains and greatly reduce the volumetric water requirements by eighty to ninety percent that would be needed to flood the land to fifteen cubits above the highest present-day mountains. But, could our present-day icecaps melt and cover all of the land in that case?

No! That is not likely. What if we add all of the water in the clouds? That will help, but we still fall far short of the volume of water required to cover the land to a point where "The water was fifteen cubits higher than the mountains which it covered" beyond the highest points of land or any hills or mountains that might have existed at that time.

The Fountains Burst Forth

Many people who are new to this topic know that it rained during the flood, but not everyone hears the other part "In the six hundredth year of the life of Noah, in the second month, in the seventeenth day of the month, all the fountains of the great deep were broken up, and the flood gates of heaven were opened."

"The fountains of the great deep" is a critically important part of the Biblical flood. There is an enormous amount of moisture and clouds in the air at any one time, but it falls very short of producing a flood that would cover even the pre-flood mountains. However, the Bible, indicating that "The fountains of the great deep were broken up", implies that there was more water to be had within the Earth somewhere beneath the surface of the ground.

There have been reports of relatively minor earthquakes in modern times where the ground opened up and water and sand shot up into the air thirty or more feet, proving that what the Bible states could occur at least to some extent. It's a matter of scale regarding modern occurrences of such an event, but "The fountains" do verifiably occasionally burst forth.

In our high-tech world of space exploration, we have images from space-probes where planets or planet moons have apparent mist being burst forth many miles into the atmosphere of the space surrounding the celestial body. Gravity of a celestial body plays a role as to how high such "fountains of the great deep" can eject water, but our modern evidence of such celestial eruptions clearly shows that it occurs on a massive scale out in the cosmos.

You could assume that this is where the extra water might have come from, but you could possibly be wrong. If the "fountains of the great deep" produced enough extra water to flood everything, then where did it go since then? We know the flood water is not all evaporated into clouds and/or on the polar icecaps because when we calculate the water volume of clouds and icecaps we are still short of the water needed to cover the greater part of the land minus the modern-day mountains. Either the water somehow flowed back into the "great deep", or we are missing some critical facts–But more on that later.

Many Extinctions

For the moment, let us assume that we have enough water here on Earth right now and that everything did flood at some point during Noah's lifetime. Does that have any effect regarding all of the extinctions we find on Earth today? There are many species of birds and animals that we find entombed in the layers of sediment, and these layers are believed to be millions of years old. Are we to believe that Noah took all of them onto the Ark? Consider the fossil that they named "Archaeopteryx". This "giant" *twenty-inch* feathered "bird-like dinosaur" is believed to be the ancestor of modern birds; it even looks like a bird. Archaeopteryx is believed to be about one-hundred-fifty million years old. Is this possible? How can we explain all of the various fossilized species found, that, according to modern geology, could not possibly have been on the boat with Noah due to the alleged age of the uppermost sediment layers, let alone available space on the Ark? As you will recall at last count, we needed about sixteen of each species with two of each kind *in every cubic foot* of the ark to get them all in the Ark at one time.

What we really need to set our focus on here is; what are the true ages of the sedimentary rock layers? We will dig a bit deeper in a later chapter, but a critical point of contention in the debate is regarding *scope* and *scale*. Just how is it that we know how old the layers are? You are encouraged to investigate this on your own; but as you do, you need to realize that most claims of the ages of rock are not as "provable" as some people would like you to believe they are. It is best not to put that specific information in this book because then people will simply repeat it as if this or that fact is true or not true. It is better on the particular question of estimating rock layer ages that each person dives very deep in to collect the data for themselves. Just beware that things are not always as someone might have them appear. The geological age determination process is similar to the species determination process. It is a matter of scope, and it is a process with a great deal of artistic license—in other words, it is a best-guess process. And, if you look into the matter for yourself, you will find that the radiometric dating is not always as accurate as it is claimed to be, and the various methods of doing so all share similar faults.

With regard to mass-extinctions and the many species that have allegedly disappeared from Earth, *time* is the key factor in determining the plausibility of any theory regarding extinct creatures found entombed in the rocky layers. The scientific evolutionary field of study contends that a massive meteorite smashed into the Earth millions of years ago creating a great amount of debris that buried many creatures and killed off the dinosaurs. This is a reasonable theory that initially does appear to pass the logic test. However, many other creatures survived, which does not reason out properly when using basic logic. Also, there are many geological layers, and any one meteor that is claimed to be a cause of extinction could only account for a couple of those layers. So logically, we are in a bit of a quandary regarding the many layers and dinosaur fossils found thus far. Additional layers are believed to be from erosion coming from mountains and volcanic ash etc. We can accept mountains and volcanoes as a possible source for any additional layers that were not deposited by a meteor colliding with Earth, but mountain and volcano sediment theory also has problems.

Since there are layers on every land-mass on Earth, and many areas do not have mountains and volcanoes from which such sediment could have eroded to be deposited in low-lying areas, how then can we explain the layers in the median elevation areas that would not have had the proper elevation to allow for that particular sediment to flow into those areas? We cannot, and that is one of the big problems with the modern long-age scientific deposition theories. We also have the problem of the fossils from creatures and trees penetrating multiple layers of sediment where each layer is believed to be many thousands and even millions of years old. The so-called "geological-column" includes the following layers: Cambrian, Ordovician, Silurian, Devonian, Carboniferous, Permian, Triassic, Jurassic, Cretaceous, and Tertiary" However, there are many more layers than those groups, and for basic information purposes, all ten geological column layer groups are not found together everywhere on earth, which basically invalidates them for use as a dating system. Whether you side with the prominent Creation view, or the prominent evolution view, one thing seems certain, and it is that both sides are wrong on many points of logic.

The many extinctions that seem to be apparent to us today cannot be properly analyzed until we have a clear understanding of the geological layers that tend to be somewhat different depending upon where you are on Earth.

Things Are Heating Up

Now let's try to connect some of these things. Continental drift is something that some Creationists foolishly deny. It is extremely obvious that certain land masses were at one point physically connected. With the advent of satellite imagery and 3D modeling, we can now study the ocean floor from the comfort of our own homes. In fact, a person today with a simple tablet computer can now intake more data in ten minutes than could the explorers-of-old in ten years, or even a hundred years for that matter. Global 3D modeling is a treasure to anyone studying geology and plate-tectonics. There are those who theorize that all land was at one time connected, and, slowly over millions of years, the land masses moved into their current positions. It is also believed that they are still in moving at a very slow rate to this very day. It is today's current rate of movement that the current movement-progression calculations are based upon. At the current rate of about two inches to a half inch per year it would take roughly 31,000 to 126,000 years to move one mile. Is this possible? Yes, it is possible. Is it likely? Not for the past. Is it logical? No, not for the past.

Many calculations have been made regarding the movement of the continents. It is the heat that is theoretically created from such movements that we suspect the volcanoes get their heat and energy from. That is a very plausible and logical analysis and is likely correct given the locations of the many volcanoes that dot the Pacific Rim. But let's look at this from a Biblical flood perspective. What would happen if instead of the land moving at about a half an inch per year, the continents actually moved into their general locations in less than a thousand years?

Since the Atlantic Ocean area shows the most obvious movement evidence and there is a distance divide almost of equal distance between North America and the Atlantic ridge and from Africa to the Atlantic ridge, from which both apparently moved. We will assume, for now, that the two continents were joined at the Atlantic ridge. This means that the greatest distance moved from current coast lines to the Atlantic ridge is roughly two thousand miles. If we assume that that two thousand miles of movement occurred in the timespan of one thousand years, the continent had to move at an average rate of two miles per year. A mile is 5280 feet, times two miles is 10,560 feet in a year, divided by 365 days in a year. This calculates to just over 29 feet per day or about 14 inches an hour which is about 1/4 inch per minute. Now while 1/4 inch is an easily observable change when you have something to index against, but it is nonetheless very slow movement. Motion itself that is that slow would not produce massive waves or cause any other serious problems once it started, especially if the motion was fairly constant.

But scientifically speaking, the idea of the land moving at a rate of 1/4 inch per minute would produce incredible heat from the continental plates sliding across their underlying foundation on the globe. However, here is something that we have to consider: If "The fountains of the great deep were broken up", then that water would have had to have come from somewhere; and since it was obviously not above the ground, it would have come from below the ground. At this point, this is sure to provoke some people to make the point that water is not that deep in the ground. But this we simply do not know, and according to the Bible's text "The fountains" were from "the great deep". The "the great deep" is a peculiar term to use if the water only came from a couple hundred feet down like the approximate depth that most modern household wells are.

Some Bible versions say the waters "burst forth", but for consistency sake we will stick with the Douay-Rheims Bible where it says "The fountains of the great deep were <u>broken up</u>". The term "broken up" is very peculiar. If there was any water beneath the continental tectonic plates then we have to ask, what allowed the land to sit atop the water to begin with? Our life-experience indicates that dirt and rock are considerably heavier than water. With rock being roughly two and a half times the weight of water, that is to say that a gallon of water is about eight pounds and a gallon of solid rock would weigh over twenty pounds; it means that, as a scientific fact, the land logically would not have floated on "the fountains of the great deep". The land would have already sunk long before any flood occurred if it was only water beneath the continents.

However, if the continental plates where resting on limited points of rocky structure combined with the buoyancy value of the rock and continental land mass, the pressures on any supporting rock structures beneath the plates would have been considerably reduced. Now, if there were supporting structures under the tectonic continental plates and those structures did not completely cover the entire space beneath the plates, then if any part of the structure were to give way and collapse, it very easily could cause a successive chain of collapsing until nearly the entire structure was resting on its global base beneath the continents.

The idea of a rocky structure being "broken up" makes sense, yet we still have to deal with any heat that would be generated from such a large mass of land moving, even if it was moving very slowly. While the friction coefficient values that the heat estimates are typically derived from are probably accurate, they are not provable in that we cannot do a lab test to move a continent. But more to the point, if there were subterranean waters that were between the tectonic continental plates and the Earth's foundational structure that lies below them, then they would not have experienced much friction therefore this would not have created the high temperatures that some estimates suggest.

If you have ever walked on a smooth wet floor, you might have experienced the reduced friction that even a thin layer of water can cause. And if your shoes are smooth on the bottom, then the slip effect is even more profound. Another example of a fluid bearing surface is found in woodworking shops or in paper handling plants, like print shops. In these type businesses they will have small air holes in the tables that the wood sheets or paper stacks are moved around on. With air injected from the surface at very low-pressure, the paper stacks weighing several hundred pounds can be moved with minimal effort without having to pick them up. You can think of this like an air-hockey table where air blowing through the tiny holes in the table's surface allows the hockey puck to glide freely across the table. This is the same effect that would occur if there was water between the tectonic plates and earth's foundational sphere. There would be minimal heat generated, and if enough heat was generated to the point where the contacting surfaces of the tectonic plates and Earth's foundation layer liquefied, then that liquid rock would also act as a liquid bearing surface thus halting the theoretical exponentially increasing temperatures from being generated beyond a certain temperature.

Now to add to this, we also have to calculate what would occur if the supporting rock structure beneath gave way. The Atlantic Ocean is just roughly three miles deep, so at the bottom of the Atlantic, the water pressure is roughly 6,500 pounds per square inch. That's a lot of pressure by any standard. North America's typical surface elevation is roughly 1000 feet above sea level. This means that the pressure of the tectonic plate at the same depth as the ocean floor is roughly 18,152 pounds per square inch. When we subtract the water pressure value of 6,500 pounds from the land pressure value of 18,152 pounds to account for land buoyancy, we are left with roughly 11,000 pounds per square inch of excess pressure at the base of the continent. In the scenario just mentioned, if the rock structure below the tectonic plate is "broken up" and collapses, then the water within that area would be ejected with a force of 11,652 pounds per square inch at the surface of the water, that is to say at sea level. To put this in perspective, a typical modern city-water-system that is functioning properly will produce about 70 psi (pounds per square) at your home. This means that the flood water would have exited the surface at a pressure that is at least 165 times that of typical city water-pressure that you experience from a garden hose.

Now, based upon basic physics calculations, water could shoot up thousands of feet into the air with pressures that high. This is consistent with the images captured by a space probe where geysers of ice are shooting into space from the surface of Enceladus, which is one of Saturn's moons. If such a phenomenon did occur here on our Earth, it would be perfectly consistent with the Biblical flood account. This also would allow for fairly quick continental movement that is far faster than the thousand-year figure we used above. This would allow for such movement to occur in a matter of a few years without boiling the oceans to nothing and evaporating them out into space. Many of the long-age continental drift supporters refuse to accept this premise because it doesn't work mathematically with their figures, but that is because they will not allow the rock-on-rock friction coefficient to be eliminated or greatly reduced and replaced with the friction coefficient when inserting a layer of water and/or molten rock into the equations.

If we calculate the total continental movement occurring in a single year's time, then the movement per minute is still only about 21 feet per minute. This is a considerably higher rate of movement. And at that high of a rate of movement it would quickly become extremely disruptive and potentially cause massive ocean surges that would oscillate across the oceans all around the world. 21 feet per minute is about a quarter of a mile per hour. A casual walk is roughly 2 to 3 miles per hour. 21 feet per minute is still quite slow, but it is a rate fast enough to disrupt things in a very substantial way.

If we can accept that the Bible indicates that there was deep water beneath the land that was "broken up", we are scientifically forced to realize that the tectonic plates would have had to have been resting on porous rock structures or rocky matrixes allowing for the "fountains of the great deep" to reside below the continental plates before the flooding began. The land would not be able to float because it is far too heavy even if much air was somehow trapped under it. If water did exist under the tectonic plates, then the plates would have to be held in place by a rocky structured network of support under most or all of the continents' base area. If the supporting rock structure began to give way, it would likely have created a successive chain of structural failures that would have gone from one coast to the other in a matter of weeks, months, years, and possibly even decades.

The plates would not be able to immediately sink because it would have taken time for the water to escape. The water would have to be ejected much like if you set a brick on a water-filled balloon and then opened the air inlet on the balloon to release the water. The water would be ejected at a rate that the air inlet size and pressure being exerted by the brick would allow–It would take time.

Now, considering that the Atlantic ridge is the obvious pre-flood connection point of the North American and African tectonic plates, we might assume that it was one such location where "the fountains of the great deep were broken up". From a Biblical flood perspective this appears to fit. If the bursting forth first occurred at the Atlantic Ridge, then the obvious direction of drift appears to be a reasonable conclusion. However, we have to consider a critical statement in the Bible where in Genesis 10:21 it says "Of Sem also, the father of all the children of Heber, the elder brother of Japheth, sons were born. The sons of Sem: Elam and Assur, and Arphaxad, and Lud, and Aram. The sons of Aram: Us and Hull, and Gether, and Mess. But Arphaxad begot Sale, of whom was born Heber. And to Heber were born two sons: the name of the one was Phaleg, because in his days the earth was divided..." This peculiar statement about the Earth being divided is often foolishly mistaken for the people being divided. The statement is a geological occurrence. It was unique, which is why it was mentioned. The name "Phaleg" holds the key to understanding the statement where it says "the name of the one was Phaleg, because in his days the earth was divided." He was named "Phaleg" "because" the "earth" was divided.

Noah to *Shem* to *Arphaxad* to *Sale* to *Heber* to *Phaleg*. This is very important in the timing of geological events if the statement "because in his days the earth was divided" is talking about the separation of the *land*. "Phaleg", or "Peleg" as it is often spelled, was born roughly one hundred years after the flood. Some people say that if the continents moved after the flood that it would have been very violent, similar to the flood fountains bursting forth. But this is not necessarily true. While it would likely have caused some very strong earthquakes and possibly caused widespread, but regionalized flooding from tsunamis, it would not have to have caused great problems. They might have had homes built, but since they had to start from scratch, their initial progress after the flood would have been fairly slow, so damages from any geological activity that occurred would not have had a great deal of impact on them like it would on our more sophisticated dwellings of today.

There is a big difference between an earthquake in a city that is populated with many modern cement structures, versus people living in stick huts or homes built with stone walls and thatched roofs or even in caves. At the time of "Phaleg", there would not have been very many people on Earth, and, logically, they would likely have lived on higher ground because they would have settled on the first ground available after the flood waters began to subside, which would obviously have been among the highest available ground at that time. The Ark allegedly came to rest on Mount Ararat. If Phaleg was born within one hundred years of the flood, then it is likely that there was only a hand full of thousands of people on Earth on the day that "Phaleg" was born. This would have changed rapidly in the years following the time around his birth due to the typical exponential nature of population growth.

It is also important to note here that just because they had enough room to once again establish life on the dirt of Earth, does not mean the that waters were completely subsided to the point where we see things today. They very possibly lived in the Ark for a generation or two. There would have been no substantial trees or other vegetation for quite a while, so they likely lived in the Ark for some time after it came to rest on Mount Ararat, but for how long that might have been is unknown. The Ark had proven itself worthy in the flood and would have been well able to handle any periodic earthquake or tsunami disruptions if any such disruptions did occur at the Ark's final resting location.

Chapter 5

Everything with the Breath of Life

Let's backup to before the flood. What possibly could have angered God so deeply that the entire world would be destroyed? Genesis Six says "And God seeing that the wickedness of men was great on the earth, and that all the thought of their heart was bent upon evil at all times, It repented him that he had made man on the earth. And being touched inwardly with sorrow of heart, He said: I will destroy man, whom I have created, from the face of the earth, from man even to beasts, from the creeping thing even to the fowls of the air, for it repents me that I have made them." What did man do?

Fallen Mankind

The Bible has fallen out of favor for so many hearts, and because of that, even more people have never even heard what it says, and when people finally do hear the key Bible stories, many of them are left confused. Often the Garden of Eden is confused with the pre-flood environment. But these two environments are not the same. The pre-flood environment would have been much as it is today, but with unrecognizable topography. When Adam and Eve where exiled from Paradise, they had to toil and till the ground just like the rest of us do in order to get their food. But at that time, it was only Adam and Eve on Earth. After being removed from the Garden, they eventually had at least five children, oddly the girls' names are not mentioned in the Bible, but they are in other ancient writings. It is believed that Adam and Eve had two sets of twins with one boy and one girl in each set. The boy from the first set was jealous of the boy from the second twin set, and according to the Bible, the older brother, named "Cain" murdered his younger brother, whose name was "Able". Then after that, Adam and Eve had another son who they named "Seth". "Seth" was wed to his dead brother Able's twin sister, and Cain was wed to his own twin sister but lived away from Adam and Eve and Seth and his wife.

Some people believe that Adam and Eve had other children but there is no formal indication of that. As the two new couples had offspring they quickly multiplied. The descendants of Seth stayed close to Adam and Eve, and they followed God's ways. It became their culture to abide by God's ways. But with Cain's family, the descendants quickly departed from Gods ways. As the story goes, there were then two distinct groups of people on Earth: The descendants of Seth, and the descendants of Cain. Cain's family left the mountain and settled below, but Seth's family remained on the mountain. Eventually, Cain's descendants lured many of the descendants of Seth down from the mountain leaving very few people remaining on the mountain.

Eventually the behavior of the Cain's descendants and those who left the mountain became so unacceptable to God that their destruction was eventually planned by God. What did they do? In the Bible it says, "And after that men began to be multiplied upon the earth, and daughters were born to them, The sons of God seeing the daughters of men, that they were fair, took to themselves wives of all which they chose. And God said: My spirit shall not remain in man forever, because he is flesh, and his days shall be a hundred and twenty years. Now giants were upon the earth in those days, because after the sons of God went in to the daughters of men, and they brought forth children, these are the mighty men of old, men of renown. And God seeing that the wickedness of men was great on the earth and that all the thoughts of their heart was bent upon evil at all times." We do not have a great deal of information on this aspect of the behavior of man, but in some ancient writing it speaks about the "giants" inferred in the Genesis Six quote just shown. There is a great deal of uncertainty regarding who the "Sons of God" were, but when they had relations with the "daughters of man", it is believed that their offspring grew abnormally large and are the "giants" spoken of early in the Bible. In other ancient writings, complimentary to the Bible, these giants eventually terrorized the population and were not easily sated as they sought people for their food, meaning that they actually ate many people who were considerably smaller than they were. At this point, we already have three huge violations: The first is that "man" was supposed to populate the Earth, not "sons of God" with "daughters of man". The second violation was that the "giant" offspring of those unions murdered people. And the third is that they ate those who they murdered, or murdered them as they ate them. Yes, this does sound a bit fantastical, but that is what is written in the ancient documents.

The ancient writings that are complimentary to the Bible mention drink houses and weapons, and all sorts of sexual deviancy. The women wanted to have sex, but did not want their bodies to be affected from having children, so they drank concoctions that would abort their babies. This was another major violation of God's Creation. Abortion has been going on for a very long time, and only when the entire culture holds to sound morals will it stop. There were all sorts of immoral debauchery at that time, all of which would violate what became the Ten Commandments that were given to Moses and the rest of the Israelites at Mount Sinai.

The rose-colored glasses that far too many preachers preach the Bible with often overlook the horrors mentioned in the Bible. This is a real sore spot with many people, because painting rosy pictures of warmth and love while overlooking the dominant theme and lessons of death murder, plundering, rape, robbery, disease, and more, in the Bible does not reconcile as "good". Many people read the Bible and they see these ugly truths and then try to reconcile these truths with the rosy picture that many preachers convey, and as explained in *Understanding the Bible - the Bible How-To Manual* AND *The Things We Don't See*, when the two views do not match, The Creator pays the price by losing those, who were once true to God, for nothing more than the inaccurate teaching of the Bible that the misguided false preachers teach.

If we are going to take the position that a Creator God exists, then we have to realize that destruction is not something that a Creator would wish to do. Logically, this means that the offences against Creation had to be utterly unacceptable for God to take such destructive action so as to flood the entire globe. When people talk about the "fall of man" or "fallen mankind" there are two points of interest. The first is the fall of man through the Original Sin of Adam and Eve, and the other is the corruption of man that occurred between the time that Adam and Eve left the Garden of Eden to when the flood occurred. "And the earth was corrupted before God, and was filled with iniquity. And when God had seen that the earth was corrupted, for all flesh had corrupted its way upon the earth..."

A Clean Slate

God was trying to start man anew with a clean slate, and, to do so, everything that man touched and all of man's evil would have to be destroyed. If God only provided regional flooding, then many evil things of the past would still have been present on the surface after the flood water subsided. If we are going to use the Bible as the base of information for this line of thought, then we have to look at some population figures. First, we have to realize that the sins mentioned in the previous section increased as time progressed, and only nearer to the time of the flood was it so bad that people were allegedly being eaten by the giant offspring, as well as women using abortive solutions to murder their babies in their own wombs for the sole purpose of women preserving their bodies in a vain youthful pre-pregnancy state. Now, regarding population size and growth, just consider a country such as the United Stated of America that in only several hundred years has grown to a population of many millions of people, even without recent immigration. Now take a similar growth rate and double or triple the allotted time for population growth that the Untied Stated of America has had and extend the lifespans of the people and now you potentially have billions of people.

If the flood occurred as stated in the Bible, then the devastation to man was no small thing. As a side note, some people might take a position and ask "How could a loving God do such a horrible thing?" This has two aspects: The first is that the people had all corrupted themselves so badly that they basically deserved death, "And the earth was corrupted before God, and was filled with iniquity." But what about innocent babies and children, did they deserve death? If the written accounts are in anyway accurate, children were also being eaten and those who were not were often offered to idols. The blood of those children would be on the hands of the parents, not God. But, if "Heaven" is a real and true place where our souls will dwell with God, then God was doing many of them a service by stopping them from sinning even more and/or from them being used as sacrifices, thus possibly saving many souls—especially children's souls from torment and/or eternal doom. Or was he damning them?

Assuming the flood occurred according to the descriptions in the Bible, and when using standard figures for population growth, the Biblical flood devastatingly changed the world by burying upwards of a billion people and all traces of them. Either the Bible is wrong, or we are missing something very big and possibly very obvious in our assessment of geology.

God's Regret

Often people claim that God is omniscient or all knowing. This seems very likely because God created everything and would then have a very good idea of actions and the results thereof. And it is likely that God understood that some souls would partake in evil, just as the angel Lucifer did. But to know the specific thoughts of everyone's hearts in advance is unlikely. A parent can know that their child might screw up now and then, but they don't expect that their children are going to be killers and rapists etc., and neither did God. But man did do evil on Earth and that is why "God seeing that the wickedness of men was great on the earth, and that all the thoughts of their heart was bent upon evil at all times, it repented God that he had made man on the earth."

The evil done by man was great, and total destruction was the only way to stop it. They were warned. The flood drowned and buried many evil people in order to start man anew through the dedicated source person–Noah, along with his family. Noah was chosen because "Noah found grace before the Lord." Noah and family had to somehow survive because God told Noah "I will bring the waters of a great flood upon the earth."

Genesis Six says that "It repented God that he had made man on the earth." This is not just because of the aforementioned sins. According to other ancient documents, man had also comingled with animals, creating all sorts of monstrosities. When we consider these things, we tend to view it from our own perspective of how life is as we know it today, but if you think back through your own life and if you are at least fifty years old you will likely recall how things changed during your own lifetime–sometimes for the better, and sometimes for the worse. God had every reason to repent of making man at that point. To put it in a personal perspective, how would you feel if your children murdered and ate people and had relations with animals and worshiped idols and sacrificed their own children to idols and aborted some of the children and dedicated others to idols and through all of their illicit behavior also produced many mutant offspring–your mutant grandchildren?

Chapter 6

Who Built the Boat?

In Genesis 6:14, God said to Noah: "The end of all flesh is come before me, the earth is filled with iniquity through them, and I will destroy them with the earth. Make thee an ark of timber planks: thou shalt make little rooms in the ark, and thou shalt pitch it within and without."

Noah was instructed to build an "Ark" sometime after Noah was five hundred years old. Genesis 5:31 says, "And Noe, when he was five hundred years old, begot Shem, Cham, and Japheth." Immediately after that, the Genesis 6:1 text says: "And after that men began to be multiplied upon the earth, and daughters were born to them, The sons of God seeing the daughters of men, that they were fair, took to themselves wives of all which they chose." This is where the wickedness of man began to proliferate. Shortly after that in Genesis 7:6 it says, "And he was six hundred years old, when the waters of the flood overflowed the earth." This means that Noah had no more than one-hundred years to complete the Ark. The question is, how did Noah know how to build an Ark?

Marine Engineering

It seems reasonable that one mechanically-minded person could build an Ark in a span of one-hundred years. But how would someone know how to do it?

For some unknown reason, in our modern culture we make foolish assumptions that we are somehow so advanced and that past civilizations spoke in an "ugha ugha" caveman-type language. However, if full-scope evolution is ever proven false as discussed in *The Science Of God Volumes 3 and 4*, then we would assume that man was created with intellect similar to God. In other words, Noah figured it out. There is a high probability that God gave Noah a far more detailed description of how the Ark was to be constructed than is briefly conveyed in the Genesis text. Some people inadvertently or subconsciously assume that the text of the Bible includes all of the words spoken at that time, but it is obviously not. Consider a book made into a movie, seldom are all of the words of a book in a movie that is based upon that book. In fact, movies typically only contain a small fraction of the words written in a book that the movie was based upon–how much more is it the case with words actually spoken from person to person?

The real question here is, was the Ark the first such craft ever built? That's unlikely. If the population growth in those first sixteen-hundred years of history is anywhere near the figures previously mentioned, then it is safe to say that within those sixteen-hundred years there could potentially have been billions of people. Many of the people born before the flood would have already been dead at the time Noah built the Ark, but most would still have been alive based upon current day population growth patterns, especially when you consider the lifespan longevity of that era. Considering the technological progress that was made in the past few hundred years of our modern era, we have to believe that they also would have had a lot of opportunity to invent great inventions before the flood. Maybe not electronic or digital technologies such as we have today, but they could easily have figured out many fundamental inventions during those sixteen-hundred pre-flood years, especially when considering the longer lifespans stated in Genesis regarding the pre-flood era.

If you consider the pyramids of Egypt, which came about after the flood, and factor in that we struggle even in our modern times to understand how the pyramids were built or to even build such monumental structures today, and then further consider that the Egyptians most likely did not have the mechanical technologies like cranes and bulldozers as we have today, you are then forced to acknowledge that they must have been cleverly ingenious people in order to accomplish such monumental tasks. The pyramids and Egyptian boats were built within only several hundred years after the flood. However, even if Noah was brilliant enough with his own ingenuity to build an Ark of any size, would he be able to build any watercraft all by himself that was as big as the Ark is said to have been?

Hired Hands

Noah could have built a boat of any size if he was ingenious enough; the question comes down to the amount of his time that it would have taken. An undertaking that was the magnitude of the Ark dimensions is no small task no matter which cubit length you choose. You might question Noah's ability to lift heavy wooden structural beams into place, but any good engineer will be able to move almost anything with the proper blocking and leverage. It might be slow progress, but it can be done. To this day we are uncertain how the large stones of the pyramids were moved in such large quantities. People of the past were by no means stupid; they had to have been quite ingenious to accomplish the marvels of the past.

A couple of key pieces of information are missing in the Bible. One piece is the ages of Noah's sons. It says "And Noah, when he was five hundred years old, begot Shem, Cham, and Japheth" Does this mean that they were triplets? Since the gestation of man is about nine months, at least two of his sons would have had to have been twins for that to be able to happen in one year if they were all born in Noah's five-hundredth year. Because of the way in which it is worded, we can assume that they were likely triplets. It's not really all that important, other than to know that they were all born in very close time-proximity and would have been too young be of much assistance until Noah was about five-hundred-fifteen and the boys were about fifteen years old, at which point the boys would have been strong enough to help their dad construct the Ark.

Since there is no specific year quantity stated that indicates, to us, the approximate time between the birth of Noah's last son and the day the flood began, we don't really know how old they were when construction of the Ark began. Based upon the Bible, what we do know is that by the time Noah was about five-hundred-fifteen years old, the boys would have been very capable. This would leave a minimum of eighty-five years for a full four-man crew to complete the Ark.

But let us not forget that there were probably many other people who were living on Earth back then who numbered in the hundreds of millions, and possibly even billions. There would have been no shortage of helpers for Noah and his sons. It is written in ancient writings that are complimentary to the Bible, that the people mocked Noah for his project and spurned his warnings about the impending doom of a cataclysmic flood. But, based upon our current life experiences, we can assume that some people probably initially took Noah very seriously. However, depending upon how long they worked on constructing the Ark with him, and the fact that the flood was so long in coming, most, if not all of those who initially headed the warning, would likely have lost interest in Noah's project and would have eventually assumed that Noah was a nutcase. In our modern era, if someone makes a future claim, then most people who don't immediately mock the claim will lose faith in only a year or two.

Thus, we can expect similar behavior from the people back then, because apparently, little has change regarding the nature of man.

If Noah did have people outside of his family helping in the early days of the construction of the Ark, then how would he have retained those helpers after the heavy mocking and ridicule began regarding building an Ark for the animals and the prophesying of a global flood?

Commerce

As we know from our own life-experiences today, you don't have to believe in something to be a part of it. So long as people are getting paid, they will typically do the job for which they are going to be paid. In our modern culture, if everyone who didn't believe in what they were doing would immediately quit their job, then our societies would quickly collapse. Money is the reason that most people show up for their jobs, and sadly, those jobs are not a labor of love for far too many people. As long as the money flows, people will work.

In a pre-flood culture, containing possibly over a billion people, some form of money would most likely have been invented and widely used. It is unlikely that they bartered with chickens and cattle etc. in exchange for Noah's labor force, although they may have. The pre-flood societies of Noah's time in those one-hundred years immediately before the flood began, were very likely well-developed societies. The likelihood that Noah had his pick of workers is very high, and it is unlikely that Noah just sat around in his rocking chair prior to starting construction of the Ark. He would have worked to sustain life just like everyone else. And since it is a common attribute for those who are dedicated to God to start up new enterprises, it is highly likely that Noah conducted some sort of business or commerce where he already had people working for him even before God told Noah about the impending doom of the flood. But of course, that point is speculation.

Since everything was wiped out in the flood, we currently have no tangible evidence beyond written accounts about some of those cultures. However, if we choose to actually search properly, then at some point, we can assume that someone will unearth treasure troves of evidence of those early cultures.

Chapter 7

Divisions

Beyond sub-continental fracturing and the subsequent continental drift mentioned in an earlier chapter, as with any era, societies arise and sometimes conflict with neighboring societies. While this is not always true, these divisions are very common between those who abide by God's guidance, versus those who deliberately thwart God's guidance. Such divisions are common in our modern era, and history shows that this has always been a point of contention. So, we can safely apply this rule to the societies of Noah's time as well.

Ancient History

In some of the extra-Biblical books that are mentioned in *Understanding the Bible - the Bible How-To Manual* AND *The Things We Don't See*, there are interesting accounts of the pre-flood era. While this information is by no means extensive, it is considerably more detailed than the Bible in that respect. If you've read the Bible's Creation account in Genesis One and/or read the earlier four volumes of *The Science Of God*, you will be

well aware of the division of "kinds" that God established when
Creating plants and creatures that would "produce seed after their
kind". When the people of the pre-flood era began to figure out
that they could alter biology regarding what we today call
"DNA", they began to pervert God's grand design and attempted
to crossbreed "kinds". This is not suggesting that they had
sophisticated labs where they altered DNA, but rather they
sought to create perversions of the existing creatures. This sort of
interaction of man and animals has always been considered
perverse and evil in the eyes of God, and according to the Bible,
such activities were punishable by death.

The people of that time had normal lives and also did normal
things, but they, too, figured out how to work with metals and
make weapons. They also figured out how to make "spirits" or
alcohol for drinking. Revelry and drunkenness became common.
Their sins were multiplied and violations of murder, theft, cruelty
and much more was a part of their everyday life.

The Watchers

In Genesis Six it says, "The sons of God seeing the daughters of men,
that they were fair, took to themselves wives of all which they chose." This is
also discussed in other ancient writings that are complimentary
to the Bible. There is debate about exactly who these "sons of God"
were. They are referred to as the "Watchers" or the "Nephilim".
Many people believe them to be fallen angels. It is from these
unions that allegedly were born the giants spoken of in a
previous chapter. Since it is not clearly indicated as to exactly
what or who they were, speculation will continue. It is written
that the watchers watched the people of Earth and found the
women of "man" to be very beautiful, at which point they decided
to come to Earth and take them as wives, thus these ungodly
relations produced mutated offspring that grew in height far
surpassing the parents' height.

In our narrow earthling view, we foolishly assume that we are the center of the Universe. And as is so subtly illustrated in big bang theology, this places our Earth in the very center of the Universe. This is a narrow view of reality. If the Universe did in fact bang into existence, then the expansion theorists may very well be correct as to our position in the Universe. However, if the Bible is correct and it is all actually created, then every planet could be considered the center of an infinite expanse. That is to say, infinity to the left and right, above and below, in front and behind–all would be equal in that infinite regard. When we look into the heavens and see the billions of stars, we then look even further and see billions of galaxies *containing* billions of stars. With the closest star to us being about four lightyears away, and since the speed of light is about six-hundred-seventy-million miles per hour and we only currently travel at under fifty thousand miles per hour, it is unlikely that we will ever travel to another star with our current understanding of the cosmos and physics.

However, since we have not found the "edge" of the Universe, we have no idea how far it goes or how many galaxies there actually are. For us to imagine that God created all of that and that we are only a single planet floating around a single star which is only one of billions and billions should be a clear indicator to us that not only is life in the cosmos probably not rare, but it is likely abundant. The likelihood that there are "aliens" in the heavens is extremely high. And knowing how the Creator works by using pattern and repetition, the likelihood that the so-called "alien" life does not exist is about zero.

Since our current human knowledge can't quite find its way to travel to any other solar system than our own, we tend to think that no other life exists, and we further assume that if it does, it could not make it here to our Earth. And if it ever did, it would have to be of very advanced intelligence. And, as is obvious in modern entertainment, these aliens are pictured as hideous creatures that are so advanced that they can build spacecraft that travel at speeds exceeding the speed of light when they often only have three difficult to navigate fingers. If they exist, then is it likely that the "aliens" are hideous creatures? No, absolutely not! As a matter of logic, if we are created "in the image of God", and if God devised all of the systems and patterns to produce life, then we can expect much the same throughout the heavens. Those patterns are seen everywhere from the smallest to the largest.

If so-called "aliens" do exist, it is likely that we would be hard-pressed to differentiate them from ourselves and they likely are not wanting to conquer our world. Nor would they have to be any more brilliant than we on Earth are in order to travel from a distant solar system. They would simply need to know how to travel at incredibly fast speeds and how to overcome the troubles that might accompany doing so. So, while it is possible that the Nephilim were fallen angels, it is also possible that since they are referred to as "sons of God", as is man of our Earth, the probability of them simply being man just like us but from other solar systems is very high. Is it far-fetched that they somehow had the ability to observe us here on our Earth? Maybe, but it is possible. If God is real, it is unlikely that all of Creation is void of life except for our own home here on Earth.

We see pattern and repetition everywhere we look, from the micro to the cosmos, so it is logical that similar life would exist throughout space. Since any other people would have been created in the same manner as Adam and Eve having been created in The Image of God, they would logically closely resemble us. However, since God did not make every planet identical and the planets likely free-formed to some extent, any habitable planets, while similar to our Earth, would have their own unique composition, it would potentially cause the overall DNA of man on those planets to differ somewhat from our own. In other words, if "alien" man does exist, then each planet with habitable life would be a Creation of "man" of its own accord, but similar to us. The books *The Science Of God Volume 3 - Day*

Five and Day Six - The Creatures - Revolution or Evolution and
*The Science Of God Volume 4 - Day Six - Evolution versus Man -
In Our Image* will give you an idea of why we might look
identical but have DNA that is not the same.

So, while we would potentially look similar, our DNA might
not be fully complimentary, which would explain the mutated
offspring produced during those pre-flood visits. Of course, this is
only logical speculation, and the "Nephilim/Watchers" could have
instead been fallen angels, but this we may never know for sure.

What we do know is that if the Bible is to be believed, then
some sort of created being, whether fallen angels or demons or
aliens, beheld the women of Earth and found them beautiful.
They then came to the women of Earth and had relations with
them, and then from those unions were borne mutated (giant)
offspring, which was not pleasing to God.

A Second Fall

The perversion of God's Creation is one of the greatest
offenses against God. Man, whether alien or Earthly, is the
pinnacle of God's Creation. Regardless of where the
"Nephilim/Watchers" were from, or exactly what they were, they
were different in some way than "man" from our Earth is, and our
Earth had only one type man with both male and female genders.
Any breeding outside of a man by men or women is punishable
by death. Thus, if the women of Earth did have relations with
any other sort of creature, it violated that primary rule of
Creation to produce seed each after their own kind.

If you recall in an earlier chapter, the crossbreeding of kinds
was mentioned. This crossbreeding of kinds included men and
animals and women and animals and woman and some sort of
alien-man or angel or demon. While man's first downfall of
Original Sin was really bad in the Garden of Eden when Eve and
then Adam consented to serve Satan by listening to Satan's
beguiling, it seems that it was actually not as bad as the second

fall of man. This would mostly be because we then knew the difference between good and evil because of Adam and Eve. God never destroyed Adam and Eve or their children. In fact, God protected Adam and Eve. God even protected Cain who killed his brother, Able. But with the people who lived during Noah's time, God found it necessary to annihilate all of them except for Noah and family. This annihilation was largely due to the perversion of God's Creation, especially the perversion of the form of man, which is to say the perversion of the DNA of man. Add to that the sins against their neighbors through murder, robbery, rape, idol worship, human sacrifice etc. and you can see that God had every reason to clean the slate in order to eradicate evil from the few remaining good people.

Chapter 8

No Permits Required

When Noah was instructed to build the Ark boat, he was told the size and nature of the boat and what it was intended for. This huge undertaking was no small structure. Such a structure in our modern day requires permits and following all sorts of regulations. Could Noah have been required to get permission for such a large structure? Probably not. While they might have had heavy populations and some regulations at that time, and they were likely fairly sophisticated, it was still a young culture compared to our forty-five-hundred-year-old cultures of today.

The Size

Just how big was this "Ark" supposed to be? As discussed in an earlier chapter. "The length of the ark shall be three hundred cubits: the breadth of it fifty cubits, and the height of it thirty cubits." This boat was longer than a football field and was to hold many animals. Now while Noah was probably a brilliant man who was able to interpret God's instructions, the boat was so massive that it was likely the largest vessel ever made to that point in time. How could someone build a wooden vessel that was possibly as long as about five-hundred feet? How could such a boat weather the storm? How did Noah cut all of the wood for the Ark and shape it and attach it together?

The Material

We may never know specifically what kind of wood was actually used for the construction of the Ark. Different Bibles translate the wood type differently. Some say "cypress", some say "gopher wood", and some simply say "timber" or "planks". Does it matter what type of wood was used? To us, no, it is indifferent. But to Noah and family and the animals, yes it mattered a great deal!

The density, strength, and flexibility of wood is very important! If the wood that the Ark was constructed of was hardwood or something like a natural spruce, it is unlikely that it would have held up as well in the flood. Woods with tight grain and are very rigid tend to fracture easily and are often much heavier. They are very durable wood, but generally are not very flexible, so when they bend they tend to fracture more readily than some softer even-grained woods do. Many softer woods that have lower density are much more flexible. Softer woods also tend to have a specific gravity that is much lower than hardwoods. That is to say that they float better—they are lighter weight, and because of that light weight they are much more buoyant.

Low-density low-specific-gravity wood is very buoyant and many of those woods when wet are very flexible. When imagining Noah and staff working with these materials, you have to take into account that they were likely working with fast-growth light-weight wood. This could allow them to handle very large lumber with relative ease.

Since they would obviously have cut a lot of lumber, we have to also realize that they were not using stone tools. A part of the second fall of man was due to the weapons that they killed each other with. They had the knowledge needed to refine and work with metals–They had tools! Noah and crew likely had what we today call "broad axes" and "adzes", and the likelihood that they had saws of some sort is also high. In addition to that, low-density woods are very easy to work with when cutting and carving them.

Other materials that may have been used are not all known because details are limited. Noah was instructed to cover the boat with "pitch" inside and out, "Make thee an Ark of timber planks: thou shalt make little rooms in the ark, and thou shalt pitch it within and without." Pitch is a thick oil tar substance possibly what we today call "crude oil". Or, pitch can be made from heating and processing tree bark. Since the construction of the Ark was a long-term construction process, likely taking decades to build, the wood laid down at the start would have rotted by the time the project was completed. However, if you had wood cut that was exposed to air and sun for a few years and then covered it "within and without" with "pitch", the oils from the pitch would soak deeply into the wood and a thick residue would remain on the outside, basically making the wood able to last many years without any rotting occurring whatsoever.

A remaining question is, did they use nails to build the Ark? If they had any metal tools, then nails were a possibility, but what kind of nails would be used and where or what on the Ark would they be used for? Did they nail the outer boards onto the structure like has been done for centuries when building modern-style hollow hulled boats?

Understanding Buoyancy

It is very important to completely understand buoyancy before trying to calculate the payload of the Ark or any such vessel. Buoyancy is based upon water weight in the case of a seafaring watercraft. Water has a weight index of about 8.3

pounds per gallon, or 1 kilogram per liter. A gallon is 231 cubic inches, and a liter is 1000 cubic centimeters.

If you take a piece of wood that is a full 1 inch thick and 5 1/2 inches wide and cut it to the length of 42 inches, it will be 231 square inches. This will make your 1-inch thick board to be 231 cubic inches in volume. Now if you weigh that piece of wood, it will almost always weight less than a gallon of water regardless of wood type. Some woods are so dense that they will weigh more than water, but those woods are few. A typical pine board the size just described will weigh about 4 pounds or roughly half the weight of the same volume of water (231 cubic inches). 4 pounds of wood divided by 8.3 of water equals 0.482, thus, the specific gravity of the wood is 0.482.

If you place this board into water, then 0.482 of its thickness will be submerged and the remainder will be above the water surface. This means that you can now place about 4 pounds of weight on the board and it will still be floating. However, as you add weight, that weight will force the board further and further into the water with each added weight that you set upon it. If the board weighs 4 pounds and you add 4 pounds of weight to it, then it will weight a total weight of 8 pounds. 8 pounds divided by 8.3 pounds equals 0.964. This means that the board will now only be out of the water about 0.036 of its thickness, or about the thickness of about 15 sheets of typical copier paper.

If you were to keep adding weight to the board, at the point that the weights and the board are identical to the weight of a gallon of water, the board will stay wherever you place it in the water, even if that happens to be somewhere under the water. But now if you add any more weight to the board it will quickly make its way to the bottom if the added weight's specific gravity is greater than that of water.

An important caution in regard to testing this basic concept is that if the wood is not painted or properly sealed in some manner (covered in pitch), then it will absorb water and its specific gravity will increase, and if that new specific gravity and the weights are added together and they exceed the specific gravity of a gallon of water (water has a value equal to 1), then your board will begin to sink. This is why Noah was to cover the Ark's wood in pitch. You will often hear that covering it in pitch was to preserve it or to stop leaks into the hull of the boat, but this is not correct. Sure it would preserve the wood, but it is unlikely that pitch was used to seal between boards. Rather, the pitch was used to seal all of the wood so that it could not absorb any water, thus keeping the specific gravity of the wood constant as it floated in the water.

Regarding buoyancy, we also have to consider that if the water did happen to have any elevated salt content, it would increase the Ark's payload due to the slight increase in the salty water's specific gravity.

Ark Buoyancy

It is important to note that dropping decimals can have an enormous impact on your final numbers; but here, for simplicity sake, we will drop some of the less-significant decimal figures. This will produce similar but differing numbers than if we include all decimals. The following numbers in this section will be close to, but different than, the numbers shown later in this book.

Now that you have an idea of how buoyancy works, we can make some rough calculations regarding the potential payload of the Ark. Buoyancy is all about the difference between the volume-weight of the liquid, versus the volume-weight of the material floating in the liquid. In our case the liquid happens to be water. Water weighs about 8.3 pound per gallon, and a gallon is 231 cubic inches. A cubic foot is 12 inches x 12 inches x 12 inches = 1728 cubic inches. 1728 cubic inches per cubic foot, divided by 231 cubic inches per gallon = 7.48 gallons per cubic foot. 7.4805 gallons times 8.3 pounds per gallon equals 62.38737 pounds per cubic foot of water. So, 62 pounds is our working water weight number for a cubic foot of water. If the Ark is 525 feet long and 87.5 feet wide, we then multiply 525 x 87.5 which equals 45,937 square feet.

If we built a pool that is 525 feet long x 87.5 feet wide and filled it to 1 foot deep with water, then that water would be 45,937 cubic feet times 62 pounds per cubic foot of water, which equals 2,848,094 pounds. That's 2.8 million pounds! If the Ark was built using the largest cubit it will be 52.5 feet high on the outside. If each of the 3 floors is 10 feet in height, it leaves us with 22.5 feet to utilize for the solid wood base of the Ark. Now we can multiply the 2,848,094 pounds per foot of water times 22.5. This means that our pool is now going to be 22.5 feet deep. 22.5 x 2,848,094 = 64,082,115 pounds of water in our 22.5 foot deep pool.

Now that we have established the weight of the water, we have to find the weight of an equal volume of wood. Assuming that the wood is one giant solid mass of wood with a specific gravity 0.482, we must multiply that by 64,082,115 pounds which equals 30,887,579 million pounds of wood. This leaves Noah with 33,194,536 of total payload, which would include the three floors. Now since heavy pine has a high specific gravity of about 0.482, we can look at a lighter material such as Northern Pine with a specific gravity of around 0.34. So now we can take the pool volume weight of water which is 64,082,115 pounds and multiply that by Northern Pine's specific gravity of 0.34 giving us an Ark-base wood-weight of 21,787,919 pounds of wood. Now take the pool water weight of 64,082,115, minus the solid Northern Pine base's weight of 21,787,919 pounds and you find that Noah has a possible maximum Gross payload of 42,294,196 pounds.

Of course, Noah has to deduct the remainder of the above water structure portion of the Ark, which is to say the walls and floors etc., but that will only be about 2,000,000 pounds, leaving Noah with about 40,000,000 pounds of available payload. However, since that 40,000,000 pound value would basically put it level with the water surface thus drowning many of the occupant animals. We must allow at least 2,000,000 pounds of tolerance for safety purposes, thus leaving Noah with about 38,000,000 remaining pounds of Net payload for animals, feed, and crew. Noah likely had well over 35 million pounds of payload weight available when the Ark was completed but was still entirely empty.

The wood that the Bible mentions in most Bibles is "Gopher wood", and since we do not truly know exactly what the properties of that wood were, we have to make some educated guesses. Since the Ark had to safely and securely carry the occupants through the flood, and God not wanting them to perish, we then can assume that "Gopher wood" was a relatively low-density wood similar to the Northern Pine used in last our calculation here. If the flood really occurred as stated in the Bible, we also have to assume that both God and Noah knew the critical needs of the project and would have wanted a somewhat flexible wood that was low density and would surface absorb the pitch in order to properly seal the wood so that the wood could not absorb any water during its post deluge voyage.

Unless Noah loaded the entire Ark with rocks, the Ark is impossible to sink, because the animals, the feed, and the people are all very close to 1.0 in specific gravity, and wood floats, so the solid-base Ark was unsinkable when fully loaded!

The Structure

Before we jump into some of the Ark size information, if you are a numbers person, it is important to note that you might choose to use a different size cubit in your calculations, and you might choose to use a different density wood than is generally used in this book, so your numbers can come out very different. However, regardless of your choices in that regard, you will find the basic premise being put forth in this book will not fail accurate math.

The Ark structure was massive, "And thus shalt thou make it: The length of the ark shall be three hundred cubits: the breadth of it fifty cubits, and the height of it thirty cubits." As mentioned in an earlier chapter, we don't know the exact "foot" equivalent of a cubit, but a "cubit" is estimated to be somewhere between twelve inches and twenty-one inches depending upon who you talk to.

While we do not know the specific cubit value regarding our modern measurements, and since flood deniers insist that we must load ten-million species on the ark, we will build our speculative Ark using a 21-inch cubit. This would make the vessel 525 feet long, 87 1/2 feet wide and 52 1/2 feet high. Let's assume that the size of the boat was 525 x 87 1/2 x 52 1/2 feet. Now when we see serious depictions of the Ark, in those pictures the Ark is always curved at the bow like modern boats, but they also draw those images with the stern being curved as well (front and back of the boat). Is this likely? Probably not.

While they likely had tools and skills to do the job, there would be no reason for them to make anything fancy on this boat—**The Ark was built <u>to serve a very specific utilitarian purpose</u>**. It was to protect the people and animals on board. The likelihood that the Ark was somewhat crude in form is very high because there is no description of anything other than utilitarian function. "Make thee an Ark of timber planks: thou shalt make little rooms in the ark, and thou shalt pitch it within and without. And thus shalt thou make it: The length of the ark shall be three hundred cubits: the breadth of it fifty cubits, and the height of it thirty cubits. Thou shalt make a window in the ark, and in a cubit shalt thou finish the top of it: and the door of the ark thou shalt set in the side: with lower, middle chambers, and third stories shalt thou make it."

Here are the details:

- an Ark of timber planks
- make little rooms in the ark
- pitch it within and without
- the length of the ark shall be three hundred cubits
- the breadth of it fifty cubits
- the height of it thirty cubits

- make a window in the ark
- and in a cubit shalt thou finish the top of it
- the door of the ark thou shalt set in the side
- lower, middle chambers, and third stories shalt thou make it

And that's about all the information we have from the Bible. As you can see from the vague description here, many Ark depictions take creative liberties that are not written in the text. Speculation is fine, but most Ark pictorial artists are not structural engineers. In order for the Ark to contain even a fraction of the kinds/species count that evolutionist insist exist today that would have to have been on the Ark while having two of each kind, the boat would have to be enormous—and very strong!

Curved bow and stern areas would require relatively thin wood in order to be bent into position. But for a wooden boat that is that large and is required to withstand the water pressure created at the base of the boat, the boards would have to be thicker than Noah could easily bend. We can imagine thin wood strips being used, but from a practical standpoint, it is unlikely that such boat building techniques were used to build the Ark. We have to keep in mind that the boat had a single utilitarian purpose, which was to take them through the storm.

When considering the construction of the Ark, we will be closer to the truth if we look at the construction methods used in older barns where they used wood pinned mortise-and-tenon construction. You can travel the world and in many places you will find barns that are well over one-hundred years old using no nails for the frame structure. The beams and supporting gussets are all wood with a pocket cut into one part and a mating protrusion out of the connected beam. Then a hole is drilled through the side of the beam with the mortise hole in it after the tenon is inserted, and a wood pin is then hammered into the hole to hold the two pieces together. This technique has stood the test of time and has proven to be very strong. Who knows, maybe these techniques have been handed down from Noah and family ever since.

Rather than relatively thin boards being nailed to a frame for the outer hollow-hull skin, as is done in modern wooden boat construction, maybe Noah did it different using little more than mortise-and-tenon type techniques for the entire structure. Let's investigate this wood-fit approach:

First let's consider the buoyancy of the materials likely used. Assuming that Noah used a lightweight wood like cypress, the specific gravity is about 0.5 or half the weight of water. This means that the wood weighs roughly half of what the same volume of water does. A gallon pail of water weighs about 8.3 pounds and a solid gallon pail size chunk of cypress wood weighs only about 4.15 pounds. Now, if we take the surface area of the boat base, it would be between 15,000 and 45,937 *square* feet depending upon which cubit length we chose to use. If the bottom of the hull was only 12 inches thick it would be between 15,000 and 45, 937 *cubic* feet. That is about 7.5 gallons of space per cubic foot. A cubic foot of water weighs about 62 pounds, times the cubic feet of the boat's base. 62 x (15,000 or 45,937) is between roughly one million and three million pounds of water. Now, since the wood has a specific gravity of 0.5, it means that a wood platform in that size range of 15,000 or 45,937 cubic feet could hold somewhere between a half-million pounds and a million-and-a-half pounds if it were only a raft-type or barge type structure one foot thick. Now, for every additional foot of base thickness beyond our original one-foot thick base we could carry upwards of a million extra pounds of cargo when using a 21-inch cubit.

Now let's take the likely approach that the Ark probably had tremendous flexibility with high structural integrity and was likely very rough on the outside. Smoothing the outer surface on the water contacting part of the logs of the boat-base would increase buoyancy due to the space being reduced between the logs. It is likely that the Ark somewhat resembled a log cabin more than a modern boat. This vessel had to enclose them safely—It did not need to be pretty. It was to be their house on a barge. They only needed to abide in the enclosed womb of the Ark just long enough for the rain and fountains to subside until they could once again walk the earth as soon as dry land appeared.

We often see depictions of a boat with a deck and a sort of house structure on the deck. But according to the text that is not part of the description "Thou shalt make a window in the ark, and in a cubit shalt thou finish the top of it: and the door of the ark thou shalt set in the side." We could take this to mean that the window was a cubit in size, but it's unlikely that is the meaning of "in a cubit shalt thou finish the top of it". The "it" is likely referring to the Ark meaning that the top/roof structure section itself was only one cubit in height. The cubit height of the roof would allow for the roof to have a slight slope of a 1:25 or a 1:50 pitch, which is just enough for water to run off and not puddle on top. It is highly unlikely that the Ark had an outdoor-style top/third deck. The roof would have been the only relatively flat surface on the Ark to see open air. This vessel was about saving the Biblical "kinds"—Noah was **not** going on vacation. The buoyancy of the base floor alone could carry a great deal of cargo, and since the boat was going to be enormous, the base was the most important part to have structurally sound. It was likely much thicker than the twelve inches used in the earlier calculation. For the Ark to be certain to not sink, it would have to be able to carry the full weight of the cargo on the base alone—without sinking.

This could easily be achieved by building a thick base as a sort of rigid raft/barge structure. The Ark was to have "lower, middle chambers, and third stories". It is possible that all three stories were used for animals, but that may not have been the case. If the base was a thick mass of wood, then the boat would not have needed to be perfectly water-tight, but the wood/logs/planks themselves would need to be well-sealed with pitch. We typically assume that Noah would have built the Ark as we construct wooden boats today, where the empty space at the bottom of the boat creates its buoyancy. However, if enough low-density wood material was used for the base structure being set forth in this book, then the base alone would be able to float many millions of pounds of cargo while water filled any open spaces between the logs or beams. Anything built above the base would not need to be very heavy or strong, even though it likely was.

If the boat's base was only six feet thick and made of cypress covered in pitch, and had a specific gravity of about 0.5, then its payload would be between three-million and nine-million pounds depending upon which cubit value we calculate with. If we use the likely in-between size that people typically use when calculating cubits, then the base would carry approximately six-million pounds of cargo. Now, since the Ark had three decks, and if the height was only thirty cubits using the mid-range figures then the total height would be forty-one feet. If we calculate the two upper floor thicknesses and the roof to be each one cubit that leaves us with the twenty-seven cubits for the three decks and the base structure. For our purposes right now, we are using the mid-range estimate of a cubit, or about sixteen and a half inches per cubit.

With about 16 inches per cubit our six-foot-thick raft-base would be 4.5 cubits. If we reserve 3 cubits for ceiling and floor construction, then with 27 remaining cubits minus the base of 4.5 cubits it leaves us with 22.5 cubits total headspace on the three decks combined. This allows us with just over ten feet of headspace per deck. Now, since some creatures such as alligators and seals etc. are land/water animals, these could not be left out of the boat or they too would perish, so they will need both wet and dry areas. To accomplish this, we can add more thickness to the raft-bottom's base floor on each end of the boat and maybe even all around the inner perimeter. If we mathematically divided the first-floor space in half and raise half of it about 3 feet, we gain over a million pounds of buoyancy giving Noah a total payload weight of over 7 million pounds using the median cubit size. However, this means that either they have to bring water in for the water/land animal kinds, or the boat needs to have water be able to flow freely in and out of the boat. Since it is unlikely that they brought water into the boat and since the boat needed to be built in such a manner so as to not be able to sink **no matter how rough the flood waters might be**, we will opt for the free flow of water in and out of the lower deck. This means that the Ark had holes in it, yet it could not sink due to the buoyancy of the pitch-covered wood.

Over the years, anyone paying attention to the various theories will have seen their share of outrageous and very questionable Ark designs. Perhaps the most ridiculous is the round Ark theory. The "round ark" theory is simple and seemingly logical design because if you place a dish in a tub of water it is very stable; however, a **round** Ark does not meet the 300 by 50 by 30-cubit specifications that God required the Ark to be built to. Other theories imply that Noah built the boat similar to how modern boats are built, and they believe he used relatively thin wood boards for this massive undertaking and then used pitch to seal up any spaces between the boards to stop any leaking into the hollow hull. This is a logical approach at first glance, however, when you consider the longest wooden boat in modern history was only about 350 feet long, plus about 100 feet of jibboom (the pointy thing sticking out the front of the ship), and considering modern era wooden ships need to be constantly bailed out to keep the water out, and that they twist and flex in rough seas and have a great deal of engineering planed into them, it is clear that this is not what Noah would have wanted to build.

The only viable and logical option for Noah to build such a massive structure that could whether the storm is for him to completely and wholly rely upon the buoyancy of the type of wood used for construction of the Ark. Allowing the free-flow of water into the boat is the only certain way to build a boat that you know for sure cannot sink no matter how bad the storm. As long as the base is properly built so as to not break apart, then the rest would not require a great deal of strength. It is important to note that the free flow of water just mentioned only needed relatively small holes to allow water into the lower deck for the amphibious creatures to swim around in. Holes large enough for water, but small enough so that the creatures won't escape or be washed out. In other words, the Ark had holes in the bottom of the base wall that would sink a normal boat, but could not possibly ever sink the Ark because the wood itself floats.

Figure 1. Ark roof has a slight angle for water run-off.

Using the mid-sized Ark calculations, Noah had about eight-million pounds of available payload. And since the high estimate of ten-million species includes about twenty-five percent that are water-only creatures, that leaves Noah with an estimated 7.5 million species. This allows just 1.06 pounds of payload per species and since there are two of each kind; it leaves us with only about one-half pound per individual creature. "Mooooooooo", that's a pretty small cow! However, we have to realize that the 7.5 million remaining species would include things like misquotes, flies, and ants etc. which offers much more room for the larger creatures.

Figure 2. Ark 3rd floor has an opening for stairs to access the second floor.

Since it is likely that the world back then was somewhat less populated than our world is today, Noah probably had his pick of trees from which to get quality lumber. There are only a few logical choices for Noah if he built a raft/barge-based boat. One way is to bundle logs together with rope and use smaller diameter logs. Or, more likely, he could have used very thick logs and used mortise-and-tenon and dovetail type techniques to hold it all together. The mortise-and-tenon and dovetail techniques offer far greater structural integrity than anything else.

Figure 3. 2nd floor has gaps on each side to feed the animals.

The logs may not have needed to be completely square, rather only clean and straight enough to provide a good sturdy fit. Keep in mind that a watertight fit is completely useless if the boat has free-flowing water for the creatures in the bottom deck. The most important part of the buoyancy construction technique is to cover every square inch of the wood with pitch so that it cannot lose any buoyancy by absorbing water, but also to have as little empty space between logs as possible.

Figure 4. Ark 1st floor has pool for amphibious creatures on far end.

As you can see, Noah using buoyant-raft/barge building techniques with a solid-core under-structure is his safest route to be certain that all of the precious cargo would survive through to the newly cleansed world. Leaks are not an issue when you rely on the buoyancy of the wood. The base-barge/raft would likely have been raw logs or hewn beams, probably each being between 50 to 150 feet in length, overlapping by half their lengths. If you have ever had the opportunity to build an all wood object using only a single pin to hold it all together you will understand what this is referring to. While it is not really a good example, wooden 3-dimensional puzzles might help a bit to better understand this concept. These puzzles usually have one piece, that when removed, the rest will easily come apart. This is not suggesting that Noah's construction would have had a single final piece to hold it all in place, but rather that properly made interlocking pieces would provide a very strong structure that was able to flex and yet not be able to easily come apart.

Imagine dovetail bevel cuts made about a third of way through the diameter or thickness of each beam/log lying flat, and then place those cuts every ten to twenty feet on both sides of the beam, but stagger them from top to bottom. Then alternate them with matching distance dovetails in each successive layer of beams. When the dovetail cuts are aligned, a properly fitted key beam can be pushed through the dovetail cuts all the way across the boat. Then pins can be driven in to hold each dovetail that holds the pieces in place. Each successive layer of logs or hewn beams would be offset or staggered by half the width of the beams, thus offering incredible strength, yet still allowing flexibility across the width. In running the beams down the length of the Ark, the goal would be to stagger the butted end so that no two log/beam ends were in the same location down the length of the Ark.

Figure 5. Contrast changed to assist in displaying the construction.
This scale model built for this book is very dark brown, almost black like pitch.

This simple technique is flexible and very durable, and depending upon the diameter of the raw trees, the entire base might have had as little as only 2 or 3 layers and potentially up to 10 layers depending upon the cubit value actually used at Noah's time. With abundant lumber and some good tools, a crew of four good men could accomplish constructing such a base in a matter of only a few years.

Chapter 9

The Family

As discussed in previous chapters, the entire world population had become corrupted and was both perverting and harming God's creatures. Since the Earth's environment was no longer untouched, the idea of creating more creatures from scratch like occurred during the Genesis One Creation events, especially man, was probably out of the question. For more about those events read *The Science of God Volumes 3 and 4*. God's Creation was good, and if not for poorly-behaved man, the flood would not have needed to occur. If only God could find one faithful family who did not fall into evil ways.

It's common to hear people saying things like, "the Bible is just a bunch of made up stories." This is obviously not true as is very apparent when you actually understand who the people in the Bible are. There are some key points that if you are not aware of, then the Bible is a lot more confusing when you read it (See *Understanding the Bible - the Bible How-To Manual* AND *The Things We Don't See*). God is a stickler for keeping basic details accurate, thus God often told people to "write this down". Recording history was and still is in the culture of those who follow God.

Noah

The line of descendancy from Adam to Noah is as follows:

- Adam
- Seth
- Enos
- Cainan
- Mahalaleel
- Jared
- Enoch
- Methuselah
- Lamech
- Noah
- Shem, Cham, and Japheth

This particular line of descendants of Adam exists for two key reasons: First is that, it is through these people that information was recorded and carried through the ages, meaning that they recorded their history. These are those who were faithful to God and stayed true to God through to their last breath. As the ancient story goes, the decedents of Cain, that's Seth's brother and is not to be confused with Cainan, departed from Adam and Eve and made their own way in life, and it is those descendants that initially began to corrupt the Earth. However, while the descendants of Seth stayed true to God, it was only those along the particular line of descendancy listed above who stayed true until the flood.

Around the time that Enoch was born, Jared had been observing the descendants of Cain who had left the mountain long before. As he observed them, Cain's descendants tried to lure Jared down to join them, and eventually he did go to investigate, but it repulsed and frightened him. When Jared returned, he warned the people to not go down to the descendants of Cain, but they did not listen. Over time, nearly everyone left the safety of the mountain, or died of old age. This all continued until Noah's day, thus leaving only Noah and family as the sole occupants of the mountain when the flood eventually occurred.

Age Limits for Passengers

Lamech was about 182 years old when Noah was born and Lamech lived to 777 years of age and died about 5 years before the flood. "And Lamech lived a hundred and eighty-two years, and begot a son. And he called his name Noe, saying: This same shall comfort us from the works and labors of our hands on the earth, which the Lord hath cursed. And Lamech lived after he begot Noe, five hundred and ninety-five years, and begot sons and daughters. And all the days of Lamech came to seven hundred and seventy-seven years, and he died. And Noe, when he was five hundred years old, begot Shem, Cham, and Japheth."

When Noah was 600 years old the flood came, and Shem Cham and Japheth were then all close to 100 years old. Depending upon how healthy he was, Lamech could have assisted Noah in building the ark for up to 95 years until Lamech's death. Also, Genesis 5 says that Lamech begot sons and daughters. It is possible that Noah had a lot of assistance in the early days of construction from Lamech's children and their children and maybe others. However, according to the ancient writings, they either died before the flood or eventually departed from the mountain. If we consider the nature of people in our modern era, we then have to acknowledge how few people have the tenacity to stick it out when things seem doubtful and are long in coming. And with the possibility of the wait being up to 100 years long, the flood was **very** long in coming.

If Noah did receive help from people other than his own children, they eventually abandoned the project or died before the flood. It may actually have initially been a labor of love for people while also having the promise of making it through the impending doom of a global flood. But a hundred years is a long time from the earliest point that God could have told Noah to build the boat. But even fifty years would be much too long of a time for impatient people to wait. While the Ark was being constructed, the remaining people were enticed by Cain's descendants and also by those that left the mountain earlier but who were of the same lineage as Noah. When you work constantly and people call on you to join them in leisure activities, it's tempting to turn away from your duties and join in the fun. So with people being people, any outside helpers likely grew weary of the project and decided to depart and join the partying and revelry down off of the mountain.

Another thing we don't know is exactly when the Ark was completed. The boat could have sat idle for years before the flood came. Let's assume that Noah had twenty strong and willing helpers for twenty years. That structure could easily have been constructed in that span of time. If the Ark did get completed early and Noah was sitting in the doorway of the boat drinking his lemonade while rocking in his rocking chair for twenty years, you can understand why people would have begun to doubt and leave to go and party in the revelry down off the mountain. They likely conceded to the point the mockers were making, which is that Noah was a nut-job.

It's the same in our modern era, people have been departing from God because the lure of the world is so great, and many people are being lied to by their pastors and priests. Many religions don't make sense due to the distortions of Biblical text that are being taught by them, and also, nothing seems to be happening as prayers appear to go unanswered (See *Understanding Prayer - Why Our Prayers Don't Work - The Prayer How-To Manual*). So, similar to our modern world that is filled with doubt, those who were not strong in faith left God and entered the world. But Noah, trusting that what God says, God does, somehow kept his family true to God because he knew that the rains would come.

A Race to Humanity

There were eight people on the Ark, Noah and his wife and their three sons and their sons' wives. Since the Bible's description says that Noah's sons were born when Noah was five-hundred years old, there could be no more than two birthing instances in that year due to the required nine-month-long human gestation period. But because there is no indication of separate birth instances, we are somewhat forced to assume that the boys were triplets. Regardless, these boys likely all physically resembled their parents as do most children. Regarding the wives of Shem, Cham, and Japheth, there is no clear indicator of their lineage.

There is some debate if the boys were triplets, because later parts of the Bible refer to the boys as younger or older brothers. But this same thing is true with Jacob and Esau who we know for sure are twins, yet they are referred to as first-born and the younger. "Sem was a hundred years old when he begot Arphaxad, two years after the flood." Some people put Shem at ninety-eight years old when the flood occurred, but we simply do not know the finer details of the amount of days between events, so things being stated in terms of years allows for a fairly large margin of error in that respect. If the boys were triplets, we have to wonder at what point in Noah's 500th year they were born, because they could have been born as late as the day before Noah's 501st birthday, which would allow the boys to be 98 years old when the flood began. We also have the issue of the term "begat", is "begat" the moment of conception, or is it the moment of birth? Unless additional information is forthcoming, we simply do not have enough information to prove much regarding whether or not the boys were triplets or their exact ages at the time the flood began.

People often wonder, "If everyone came from Noah, then where did the different races come from?" This same point is true if we all came from Adam and Eve. Various people claim that the quantity of races runs between three and six different races. However, if you examine these races, some of them are very similar and are broken up by geographic area, yet they share some very distinct features with people from other areas. If we consider Noah's sons' families, we see that they each had a wife and it is likely that those wives were from different families or clans. When we examine people on Earth, we see three distinct races with populations large enough to warrant making notable distinctions, and they are: Asian, African/Black, and White skin. Most other races are likely a combination of those three. Defining what a "race" is, is much like defining species when looking at this in the same way that evolutionary species are defined in our modern era.

It is likely that the three major races are descended from the three sons of Noah, with the major differentiating attributes being contributed by the boys' wives. While we have a pretty good idea of the migration of their families, that trail is not specifically known for all three. What we do know with relative certainty is that Cham and his wife are the ancestors of the Native Africa people, and Shem and his wife are the ancestors of the European Native White people, leaving us to logically conclude that Japheth and his wife are the ancestors of the Native Asian people. From there the distinctions in people are due to closely aligned breeding within tribes causing distinctions to become readily discernable because people very possibly likely only reproduced within each of the sons' own tribes in those early days.

Since their wives were likely from different families, their offspring would have followed the pattern of those bloodlines combined with Noah's bloodline, thus possibly resulting in the three major current-day races. After years of separate clans who were breeding only within each clan, distinct traits were enhanced over time. Those distinctive groups interbred and offered a more distinctive variation. This is the same thing that occurs when we crossbreed animals. We change their look, yet they are still dogs or cats. This is the only part of evolution that is accurate, and it ends within that scope, because we are all people and can interbreed with any other races or sub races on Earth while producing beautiful offspring that show clear traces of the attributes of the parents.

If the Bible is a true account, and the flood actually occurred as described, we then have to accept that Noah had three sons and those three sons had wives as indicated in the text. And when the flood ended the three couples each began their own family and likely passed along strong traits of the mothers. We witness this very quickly with animals because their birth to reproduction maturity and their gestation periods are so short, thus allowing us to influence traits by limiting a dog's or a cat's access to breeding partners—this is no different with man. As the three couples' families began to grow, they developed separate clans within the larger tribe and as those populations grew, and sub-clans formed. It is these newly developed sub-clans that are the readily apparent and likely sub-races that we see today.

You will notice in the evolutionary line of thinking that the white people are implied to be "further evolved" than the other races, but this is not true—all races are equal. Some races might be more representative to the actual origin-parents because they stayed more within their own larger tribe and mixed well within that tribe—but, all races are equal.

The full-scope primate-to-man evolutionists believe that "Africa is the cradle of civilization", but this is not true. They say this because DNA tracing shows that the African people have a limited range of DNA and the DNA traces from out of Africa to the Middle East area, and from there the DNA trail is spread abroad. However, this bigoted thinking is foolish and should be taken as an insult to every true African. Man did not evolve from primates, see *The Science Of God Volume 4 - Day Six - Evolution versus Man - In Our Image*, plus, the Ark that Noah built landed somewhere in what we today call the Middle East/European area.

The Bible indicates that Cham's decedents went to Africa. Since Africa is somewhat isolated by water, they had no place to expand, which would explain why their DNA trail is more pure with less mixed-race DNA. The DNA of Cham and his wife was geographically isolated. Shem and his wife stayed closer to home around where the Ark rested after the flood, and Japheth went off his own way. The mix between Japheth and Shem is far more likely and is due to that fact that the land masses are more contiguous. Contiguous land mass offered greater diversity in the clans' long-term breeding selection as they easily migrated to-and-fro. The implied Biblical migration of Noah's sons and their clans and sub-clans matches the modern-day DNA migration evidence, as does the isolated nature of the African continent and its inhabitants, although some of that data is read backwards.

Chapter 10

Promises Made

People often ramble on about the "covenants" spoken of in the Bible, without truly understanding exactly what was being covenanted. These "covenant" promises from God go back to Adam and Eve. As each faithful child from the subsequent generation found his way to God and served God, God would promise that the Salvation from Adam and Eve's Original Sin that was originally promised to Adam and Eve and their righteous offspring, would come through the current dedicated child—And that is the primary and most important covenant ever made with man.

Many other covenants were also made that would narrow the field of candidate families through whom Salvation would be borne. Here is an example of the dedication that would cause God to look favorably on such people beginning in Genesis Six:

"Go out of the ark, thou and thy wife, thy sons, and the wives of thy sons with thee. All living things that are with thee of all flesh, as well in fowls as in beasts, and all creeping things that creep upon the earth, bring out with thee, and go ye upon the earth: increase and multiply upon it. So Noe went out, he and his sons: his wife, and the wives of his sons with him. And all living things, and cattle, and creeping things that creep upon the earth, according to their kinds, went out of the ark. And Noe built an altar unto the Lord: and taking of all cattle and fowls that were clean, offered holocausts upon the altar. And the Lord smelled a sweet savor, and said: I will no more curse the earth for the sake of man: for the imagination and thought of man's heart are prone to evil from his youth: therefore I will no more destroy every living soul as I have done. All the days of the earth, seedtime and harvest, cold and heat, summer and winter, night and day, shall not cease."

"And God blessed Noe and his sons. And he said to them: Increase and multiply, and fill the earth. And let the fear and dread of you be upon all the beasts of the earth, and upon all the fowls of the air, and all that move upon the earth: all the fishes of the sea are delivered into your hand. And everything that moves and lives shall be meat for you: even as the green herbs have I delivered them all to you: Except that flesh with blood you shall not eat. For I will require the blood of your lives at the hand of every beast, and at the hand of man, at the hand of every man, and of his brother, will I require the life of man. Whosoever shall shed man's blood, his blood shall be shed: for man was made to the image of God. But increase you and multiply, and go upon the earth, and fill it. Thus also said God to Noe, and to his sons with him, Behold I will establish my covenant with you, and with your seed after you: And with every living soul that is with you, as well in all birds as in cattle and beasts of the earth, that are come forth out of the ark, and in all the beasts of the earth."

"I will establish my covenant with you, and all flesh shall be no more destroyed with the waters of a flood, neither shall there be from henceforth a flood to waste the earth. And God said: This is the sign of the covenant which I give between me and you, and to every living soul that is with you, for perpetual generations. I will set my bow in the clouds, and it shall be the sign of a covenant between me, and between the earth. And when I shall cover the sky with clouds, my bow shall appear in the clouds: And I will remember my covenant with you, and with every living soul that bears flesh: and there shall no more be waters of a flood to destroy all flesh. And the bow shall be in the clouds, and I shall see it, and shall remember the everlasting covenant, that was made between God and every living soul of all flesh which is upon the earth. And God said to Noe: This shall be the sign of the covenant which I have established between me and all flesh upon the earth."

Promises Kept

The covenant made in Genesis Seven was a long-term promise to man that God would never again wipe out all flesh with a global flood no matter how terribly we behave. However, this does not bar God from dealing with our horrendous behavior in other destructive ways. But, let's back things up a bit and go back to before the flood.

"It repented him that he had made man on the earth. And being touched inwardly with sorrow of heart, He said: I will destroy man, whom I have created, from the face of the earth, from man even to beasts, from the creeping thing even to the fowls of the air, for it repents me that I have made them. But Noe found grace before the Lord. These are the generations of Noe: Noe was a just and perfect man in his generations, he walked with God. And he begot three sons, Shem, Cham, and Japheth. And the earth was corrupted before God, and was filled with iniquity. And when God had seen that the earth was corrupted (for all flesh had corrupted its way upon the earth,) He said to Noe: The end of all flesh is come before me, the earth is filled with iniquity through them, and I will destroy them with the earth. Make thee an Ark of timber planks: thou shalt make little rooms in the ark, and thou shalt pitch it within and without. And thus shalt thou make it: The length of the ark shall be three hundred cubits: the breadth of it fifty cubits, and the height of it thirty cubits. Thou shalt make a window in the ark, and in a cubit shalt thou finish the top of it: and the door of the ark thou shalt set in the side: with lower, middle chambers and third stories shalt thou make it. Behold I will bring the waters of a great flood upon the earth, to destroy all flesh, wherein is the breath of life, under heaven. All things that are in the earth shall be consumed. And I will establish my covenant with thee, and thou shalt enter into the ark, thou and thy sons, and thy wife, and the wives of thy sons with thee. And of every living creature of all flesh, thou shalt bring two of a sort into the ark so that they may live with thee: of the male sex, and the female. Of fowls according to their kind and of beasts in their kind, and of everything that creeps on the earth according to its kind; two of every sort shall go in with thee that they may live. Thou shalt take unto thee of all food that may be eaten, and thou shalt lay it up with thee: and it shall be food for thee and them. And Noe did all things which God commanded him."

This section of text is both a command and a promise. First God takes note of the extensive corruption and wickedness occurring, and then God proceeds to command Noah to build and prepare an Ark, after which God sums it up with "two of every sort shall go in with thee that they may live" This tells Noah, in no uncertain terms, that everything is going to be wiped out including him if it is not on board the Ark. The two key implied promises are that you and the creatures will be saved, and that it's going to rain.

Looks Like Rain

There have been many theories offered regarding the flood water and where it came from. There are many who use the rainbow as a type of evidence that there was a heavy canopy of water suspended like clouds and the environment was more of a tropical paradise. This view is based on two key points: One often injected is that it was somewhat like Eden in Noah's time and the other is that, since God was going to place a bow in the sky and we understand how rainbows are made, then a rainbow could not previously have been visible at that time when they say that there was a canopy of clouds blocking the Sun from being able to cause a rainbow to appear from Noah's Earthly perspective. While this sounds like a good explanation, it does lack credibility.

There are many problems with a "canopy" of mist or clouds or water because it would eventually likely either have boiled them out, or frozen them. The canopy-theory clouds would have been so thick and dark to contain even a small portion of the flood waters that it would have been very dark all over Earth during the entire pre-flood era. If that were the case it would have been so dark that the entire era would have been darker than nighttime. And further, the people would have been hard-pressed to track time in that unlikely constantly-cloudy theory because God said "Let there be lights made in the firmament of heaven, to divide the day and the night, and let them be for signs, and for seasons, and for days and years." This Genesis One text is clearly indicating that these "lights" would be able to mark "signs, and for seasons, and for days and years", logically speaking, what is the point if the entire Earth is constantly overcast?

Another reason this canopy theory is used is to attempt to explain where enough water came from to flood the entire globe over the mountains' height, which was partly addressed in an earlier chapter. But it, too, faces the same problem of explaining where all of the water disappeared to afterwards. In addition to that, the canopy theory is also derived from mistranslations of Genesis One as discussed in *The Science Of God Volume 1 - The First Four Days*. When translations use an earthly perspective, they tend to change words in the text such as changing "firmament" to "canopy" or "dome" or "clouds", thus allowing utter misunderstanding of the true message in the text. It is unlikely that the climate and skies of Noah's time were any different than what we experience in our modern times forty-five hundred trips around the Sun later.

It is unlikely that the rain came the day Noah and crew completed the Ark. They probably had several years to prepare and plan and stock up on essentials while they waited a number of years after construction was complete before the flood began. If the culture of that time was anything like our modern era, then Noah and family were likely mocked for staying true to the commands and promises from God.

Here's something to ask yourself: Do you think it ever rained before the flood? Some people claim that it never rained because everything was watered from beneath according to Genesis Two "...in the day that the Lord God made the heaven and the earth: every plant of the field before it sprung up in the earth, and every herb of the ground before it grew: for the Lord God had not rained upon the earth; and there was not a man to till the earth. But a spring rose out of the earth, watering all the surface of the earth."

In a King Jams Bible Version it says "But there went up a <u>mist</u> from the earth, and watered the whole face of the ground." Rather than "But a <u>spring</u> rose out of the earth, watering all the surface of the earth."

If you read the text without any late post-Reformation translation influence, it is clear that "before it grew" was stating that the Earth had either bubbling waters or mist or both coming from the ground. If we look at this with a scientific perspective, the Earth was likely very warm when it first came to be formed, and there would have been steam or fog rising up from the ground while the plants and atmosphere were being developed. This view supports the canopy idea somewhat, but since that view is dependent on water bubbling up rather than a mist, it would logically defeat itself because, from our scientific view, the mist would have been caused due to the warmth and would likely have been a warm ground-fog like you see when plowed fields are warmed by the Sun and the air cools down where you will see steam or fog rising from the ground. But at the point of Creation, the ground-fog would have been much heavier. Such a fog would have made life very difficult for man because depending upon the density of this "mist" you would not be able to see around you. And if the fog was very low you might not be able to see the ground more than a few yards in front of you.

After Creation was complete, there would have been a period of acclimation for everything to come into balance before Adam arose from the dirt-slime. While we can imagine that there was a canopy and/or water misting up from the ground, it is unlikely that this would have been the case over sixteen-hundred years after Adam was created. "for the Lord God had not rained upon the earth" is strong implication that it was going to rain before the flood was ever mentioned or even conceived. There is a lot of information packed into Genesis, so we must use common sense and fundamental scientific observation when attempting to understand the text. The rose-colored glasses view that so many people understand the Bible through, causes us to invent things that are not logical and have more of a fairytale nature about them. The Sun did shine before the flood for Noah and family, and they likely eagerly, but with great reserve, waited while watching the rain clouds come and go.

Problems Bubbling Up

Based upon our human nature and the fact that we typically understand things from our earthly experience and earthly perspective, Noah and his family almost certainly expected the flood to come through heavy rain without any thought of other options, but little did Noah know what God had in store for the world. God said "Behold I will bring the waters of a great flood upon the earth, to destroy all flesh, wherein is the breath of life, under heaven. All things that are in the earth shall be consumed." Based upon this brief statement, Noah would have had no reason to expect anything other than rain, but with God thinking outside of the box, things were not as expected.

"For yet a while, and after seven days, I will rain upon the earth forty days and forty nights; and I will destroy every substance that I have made, from the face of the earth. And Noe did all things which the Lord had commanded him. And he was six hundred years old, when the waters of the flood overflowed the earth. And Noe went in and his sons, his wife and the wives of his sons with him into the ark, because of the waters of the flood. And of beasts clean and unclean, and of fowls, and of everything that moves upon the earth, Two and two went to Noe into the ark, male and female, as the Lord had commanded Noe. And after the seven days were passed, the waters of the flood overflowed the earth. In the six hundredth year of the life of Noe, in the second month, in the seventeenth day of the month, all the fountains of the great deep were broken up, and the flood gates of heaven were opened: And the rain fell upon the earth forty days and forty nights. In the selfsame day Noe, and Shem, and Cham, and Japheth his sons: his wife, and the three wives of his sons with them, went into the ark: They and every beast according to its kind, and all the cattle in their kind, and everything that moves upon the earth according to its kind, and every fowl according to its kind, all birds, and all that fly, Went in to Noe into the ark, two and two of all flesh, wherein was the breath of life. And they that went in, went in male and female of all flesh, as God had commanded him: and the Lord shut him in on the outside."

This text states that Noah and his family and all the animals that God sent were locked in the Ark a full week before "the waters of the flood overflowed the earth". The next event is typically shown in movies as an explosive event for entertainment purposes. "all the fountains of the great deep were broken up, and the flood gates of heaven were opened" When the "fountains of the great deep" are depicted, it is often done as if suddenly they burst forth and exploded into the air with tremendous force. While this is possibly true, it is not necessarily how it began near Noah and family. If that much water came from beneath the surface of the land it wasn't from a handful of relatively shallow aquifers, it would have been from the "great deep".

As was discussed in an earlier chapter, the only logical answer that will explain continental drift and enough waters coming from the "great deep" is if the "great deep" was actually beneath the continental plates. Any other thought is simply irrational based upon the text and the physical evidence we see on the ocean floor today. There most likely was great turmoil during the flood, but that does not mean that it began violently from the first moments at the specific location where the Ark was built.

We might imagine that there would have been thousands of people in an angry mob trying to get into the ark the moment the first drop of rain fell, but that is unlikely. The flood took forty days to completely cover the highest hills by fifteen cubits. The first few days, people might have assumed it was just heavy rain as normal. That is, unless of course the fountains of the great deep did explosively erupt. In either case, few people would have tried to enter the Ark. If the great deep did burst forth explosively, then things would have been quickly washed away, and if it began to rain more slowly the first day or two people would have assumed it was just another rain. But when you realize how in only a matter of minutes flash-flooding occurs with torrents of water, you can then see very quickly how the people would have been without options, at which point they would have been hanging on for life as Noah and family rested safely in the Ark. Additionally the water coming from the mountain run-off would have indeed caused much flash flooding in the areas approaching the base of the mountain that Noah was on, making it very difficult for anyone having second thoughts to make their way to Noah.

This is where aspects of the Ark get very interesting. Almost every depiction of the Ark weathering the storm depicts violent seas during the rains. And while this might have been so, it does not mean that Noah and family ever experienced any rough waters at any point during the flood. We will dig into another part of this in a later chapter, but regarding the forty days of rain and fountains of the great deep, Noah and family may have only begun to float in the Ark starting the last couple of days of the forty days of rain.

As you will see as you continue reading, we really do have our perspective of the flood event incorrectly viewed using our relatively recent understanding of the Earth's topography. At this point, we have to come to the understanding that the topography of the world that Noah left behind would be absolutely and completely, and without question, unrecognizable to us today. The mountains we have today did not exist then as is clearly evident and as was discussed in an earlier chapter. And most, if not all, mountains or hills from before the flood were most likely completely destroyed during the flood. The Atlantic Ocean did not exist, and what we call the Pacific Ocean was probably considerably larger since the continents drifted towards it from both east and west and even possibly did not exist as it could also have been land above water that sunk when the "fountains of the great deep burst forth".

Because there is no other option for the "great deep" than for it to be very deep at a level deep enough to be below the base of the continental tectonic plates, we can logically know that there would have been many earthquakes as the supporting structure beneath the plates began to crumble. If the water was not from beneath the continental plates, then the flood account in the Bible is simply false. This whole process took forty days and there is no indication of acceleration of change in intensity. What we do know is that if it had been only rain falling all around the globe for forty consecutive days, the clouds would have been depleted of rain long before the forty days ended. But, as indicated in the text, if we allowed the fountains of the "great deep" to "burst forth" to assist in the flooding process, then the flood account becomes far more logical and far more believable. First, the water could easily shoot up to, or even above, typical modern-day cloud heights. This would allow for two things: first it would have to come back down in the form of rain, and second, with the massive crushing activity and the pressure at which the water exited the "great deep", the water in places would have been heated to a point of vaporizing like when taking a hot shower. Then as the water condensed it would have been able to form clouds and rain back down. This is all basic observational type science that we can reproduce in miniature in any of our experiments regarding such.

The only way it could rain for forty consecutive days would be if water was quickly added to the atmosphere. There is also the added possibility that waterspouts could have developed lending to the amount of rain falling around the globe, but that would not have had any effect on the final water depth after the flood waters ceased to continue because the water from a waterspout is simply recirculating the water. So, while water spouts could have made it rain harder, it would not have produced more water.

Now let's assume that this upheaval produces tsunami waves. Tsunami waves have been known to travel at speeds up to six-hundred miles per hour. At that speed it would take forty hours for a single wave to make its way around the entire globe. If the edge of a continent began to collapse it would set off a domino-type chain reaction of catastrophic proportions. We will assume that most of the continental plates were supported by a somewhat fragile porous rocky structure, possibly similar to lava-rock, since most continents appear to have shifted a substantial distance based on current-day evidence seen on the ocean floors.

If the edge of a single continent suddenly had its supporting structure crumble beneath it, it would send a wave so massive unlike anything that had ever before occurred in the history of Earth and also unlike anything that occurred after the cataclysmic effects of the flood finally ceased. These tsunami waves would vary in size depending upon the rate of failure of the supporting porous-rock structure beneath the continents. If all or most continents shared a similar fragile supporting structure, then a several-hundred-foot tsunami slamming into another continent could trigger the other continents' collapse.

This would occur for a few reasons; the first is the force that the water waves would hit the edge of the continent with would be extremely powerful. Plus we have the momentum of the water dragging across the surface of the receiving continents, And finally, there is the suddenly added weight of the water that was deposited onto the receiving continent, and when all added together this would cause enough force to return in like kind to the originating continent, thus causing the originating continent to further collapse and then repeating the same oscillating cycle until the subterranean waters were substantially depleted and the continental plates came to rest on the foundational surface of the globe.

Experience tells us that the distance a tsunami can travel inland is somewhat limited. However, that is really dependent upon the topography and the size and speed of the wave. A relatively small ripple wave of maybe one hundred feet in height going about fifty miles per hour would travel only as far as the volume of water can be spread. However, a wave having the power of a wall of water traveling at hundreds of miles per hour and having a cross-section of thousands of feet of breadth and a height of several hundred feet could travel inland across a relatively flat topography for hundreds of miles. As the subterranean structure collapses with the impact from each successive tsunami, the waves would be able to make it further and further inland as the ocean was rising, plus the land was dropping from the crumbling structure beneath it. The dynamics on this are complicated to try to untangle because there is a lot more to it all, too much to put in a book about the viability of the bigger picture of a worldwide flood and the survival of animals etc.

If you take the time to really examine the maps of the ocean floor, you will quickly notice the movement lines where traces of the continents had been deposited due to the turmoil and movement that occurred. Many of these lines defy logic, yet they are still there. If you deny their existence then you can invent any imaginative theory about plate tectonics that you wish, but those markers tell a story of geological activity consistent with the Genesis flood account.

A part of the point of this section is that the excess water had to come from somewhere, and according to the Bible's account of the flood; it was from the "great deep". Another key point that this has all been building to, is that at one point, it is likely that there was one giant continent. This view is based upon the complex movement trails shown on detailed maps of the ocean floor. Depending upon where the first supporting structure began to give way, Noah and family could have been sitting safely and firmly on the ground for most of the duration of the forty days of flooding.

Based upon our buoyant barge-raft-style foundation for the Ark as described in an earlier chapter, the Ark fully loaded would have had a bottom depth below water level of about eight cubits, and if the Ark had any sort of a keel made of a higher density wood for stability and durability, it could easily add a cubit of depth. This would bring the depth to nine cubits. The text says that "all the high mountains under the whole heaven were covered. The water was fifteen cubits higher than the mountains which it covered." This means that the Ark could have rested undisturbed for the majority of the flooding period of forty days, even until the very last moments of the forty days.

Of course, we have to assume that the Ark was on one of the highest elevation areas that existed at that time. If the Ark was on one of the highest places, then Noah and family could have still been sitting firmly on the ground when *everything* else was covered in water, and then finally only being lifted the six or seven remaining cubits the last several days and possibly less than several days. We'll get deeper into the timing of the Ark beginning to float in a later chapter. The point here is that the Ark may not have experienced any rough seas during the forty days of rain, because it may have been the last thing on earth sitting on ground above water just before it eventually began to have water deep enough beneath it to float it up off of the ground.

Let's also take a look at the tsunamis that would have occurred oscillating across the oceans for over a month. If you study the global ocean floor maps, it is evident that the general area that the Ark allegedly was built on would have been somewhat centrally located on the massive continent based upon the topological trails. Since the tsunami waves are limited in their ability to travel over land, those waves would not have immediately made it to Noah's location. This would have allowed the Ark to sit safely out of harm's way in the early part of the flooding period. Then we have the additional factor of an increasingly depleted water source beneath the continents along with increased water depth above land. These two factors would have done one thing from two angles. The first is that, as the continent was slowly sinking and coming to rest on the foundational surface of the globe, the rate of ejected water would be greatly reduced thus causing increasingly calmer rising waters. Also, any fountains from the great deep would have ever-increasing resistance as the relative water level was rising, resulting in reducing the pressure and motion of the water; this means that the water may have been much more gently bubbling up in the later part of the forty days of flow from the fountains of "the great deep".

Noah and family could have had a very gentle ride all through the forty days of rain. But since they were afloat for about a year they may have encountered severe weather.

Chapter 11

All Aboard

Before the rains of the flood began "In the selfsame day Noe, and Shem, and Cham, and Japheth his sons: his wife, and the three wives of his sons with them, went into the ark: They and every beast according to its kind, and all the cattle in their kind, and everything that moves upon the earth according to its kind, and every fowl according to its kind, all birds, and all that fly, Went in to Noe into the ark, two and two of all flesh, wherein was the breath of life. And they that went in, went in male and female of all flesh, as God had commanded him: and the Lord shut him in on the outside."

Will Everything Fit?

When debating this topic, Creation supporters will claim that the Ark had plenty of room to get two of every species aboard the Ark, but evolution supporters refute that with the claims of trying to get two of each kind of *millions of species* on board, including a pair of one-hundred-thirty-foot long "Argentino-saurus" and a pair of seventy-five-foot long "Brontosaurus" etc. This is where things get really interesting. If the evolutionists get their way, then it is utterly impossible to get even a small fraction of the millions species on the Ark.

First, let's approach this by categorizing species. Species determination is an art-form more than it is a science. Many, actually most, of the species included in the 7.5 million species the evolutionists require to be on the Ark are small... very small! Not cricket and parakeet small, but micro small, the kind that do not have to be considered because they live on the animals. This vast array of micro creatures that are already living on the other creatures, or are naturally on most everything, reduces our cargo candidates to somewhere around 1.5 million species, half of which are insect creatures and not what we think of as "animals". This immediately reduces our larger species to about 750,000, yet we need two of each so we are back up to 1.5 million individual creatures. Since there is an insistence that we must pack that many creatures into the Ark and our initial capacity figures were based on a smaller Ark volume, we will increase the volume using the cubit size used for a 525-foot Ark size described in Chapter 8.

Using a 21-inch cubit we will calculate each of the 3 Ark decks with the following cubits 3 x 300 x 50 = 45,000 square cubits. But to convert to feet and meters, first, we will first take the basic Ark dimensions of 300 x 30 x 50 and convert those to feet and meters. 21 inches per cubit divided by 12 inches per foot equals 1.75.

Converting 21 inch Cubits to feet:

300 cubits long x 1.75 = 525 feet long

50 cubits wide x 1.75 = 87.5 feet wide

30 cubits in height x 1.75 = 52.5 tall

525 x 87.5 = 45,937.5 square feet per floor

3 floors x 45,937.5sq ft. = 137,812.5 sq feet of total floor space.

Allowing each floor the headroom of 10 feet offers Noah 1,378,125 total cubic feet of cargo space. Calculating with two of each of the 750,000 creatures, being 1,500,000, allows an average space of 0.91875 cubic feet per creature, or 1.8375 cubic feet per creature couple.

When converting 0.5334 meter cubit to meters:

300 cubits long x 0.5334 = 160 meters long

50 cubits wide x 0.5334 = 26.7 meters wide

30 cubits in height x 0.5334 = 16 meters tall

160 x 26.7 = 4272 square meters of floor space per floor

3 floors x 4272 square meters = 12,816 square meters of total floor space.

Allowing each floor headroom of 3 meters offers Noah 38,448 cubic meters of total cargo space. Calculating with two of each of the 750,000 creatures, being 1,500,000 total, allows an average space of 0.51264 cubic meters per creature, or 0.102528 cubic meters per creature couple.

The above figures leave us with available room for a base that is 22.5 feet thick (or about 7 meters).

Just over 1.8 cubic feet per creature-couple is about the same as 13 1/2 gallons of water worth of space, or about the size of a 15-gallon aquarium. This is considerably better than the original 44 creatures per cubic foot mentioned in Chapter 2.

But this is still not realistic, after all, we still have to mathematically deal with cattle and dinosaurs etc. To compensate for some of that, we have to acknowledge that very many of the species are birds and mice etc. When considering creatures such as mouse-like rodents or "myomorpha", there are said to be over eleven-hundred species. With each of the 1,524 mouse-size species and those couples having a shared space of 1.8 cubic feet, the 1,524 creature couples would use 2257 cubic feet of total space (63.9 cubic meters). If we put these 3,048 little critters in to 220 cubic feet of space (6.2 cubic meters), then it would free up 2,037 cubic feet of space (57.7 cubic meters). A typical modern bedroom is about 11 x 11 x 8 feet or 968 cubic feet (27.4 cubic meters). This means from the mouse species alone, we have gained the space of two typical size bedrooms.

The Fish and Amphibians

We already eliminated the water-only habitat creatures because they are all safe in the water. However, we do have to deal with the amphibians. The amphibian situation is similar to the mice, where small ones like salamanders and lizards could be grouped into far smaller confines to allow much more room for the larger creatures. Using the same calculations as with the mice, there are over six hundred species of salamanders and over four-thousand species of lizards. You can go down a long list of any group and you will find the same pattern emerge with tiny creatures. In other words, most species couples need far less than a cubic foot to live.

Some people could claim that it is cruel to have any creature so tightly confined, but if the choice is death versus being trapped in your bedroom for a year, which will you choose? The Ark was not a honeymoon cruise; it was about survival, so any creature discomfort was a third level issue. First is to get on the Ark, second is to be fed to survive the entire duration of the voyage, and then only after those do we care about creature comfort.

The Birds

Now, granted, the Ark would be very full with 1.5 million creatures, yet it is highly likely that some creatures roamed freely within a given space or even possibly within nearly the entire vessel. Depending upon the size of creature, this could eliminate many birds from the equation altogether, because the birds would be able to move freely while perching on the larger animals or roosting in corners and on beams etc. Since it is claimed that birds have about 11,000 species that's about 22,000 individual birds having two of each kind included, which frees up about 19,800 cubic feet of space (560 cubic meters) for all of those creature couples, divided by a bedroom space of 968 cubic feet (27 cubic meters) gains us over 20 typical bedroom sized spaces.

You will see that when you go through the various species, this pattern will repeat again and again. However, since it is claimed that birds and dinosaurs are related we still have to account for the alleged bird cousins that can range from only inches in length to 130 feet in length. And with 700 different species, that becomes a problem real quick. However, in the unlikely event that the dinosaurs and birds do share the exact same ancestry, we will still classify dinosaurs for our purposes here as "beasts" or animals because they certainly are not going to be flying around freely in the Ark, or flying at all for that matter.

The Animals

Since there are over 700, and possibly over 1000, dinosaur species that range from under 1 pound up to about 130,000 pounds we will use the average of about 4,200 pounds per species. This is the size of a smaller species of elephant or a rhinoceros.

To begin to grasp just a sample portion of the species of land animals, take a look at these species' quantity estimates of the various kinds of animals listed below:

- Bear: 9 or 8
- Horse: unknown, because they are referred to as "breeds"
- Cat: 40 or 60
- Dog: unknown, because they are referred to as "breeds"
- Gopher: 13 or 35
- Muskrats: 16 or 142
- Rabbit: 29
- Skunk: 12
- Deer: 43
- Monkey: 160 or 200
- Gorilla: 2
- Ape: 20 or 26
- Orangutan: 3
- Giraffe: 4

As you can see by this very brief list of creatures, the species determination is flexible based up the ranges shown. Many evolution enthusiasts will insist that species determination is a refined and exact science, but by asking only a few pointed questions, you will quickly prove that the practice of species designation is an *art* rather than a *science*. But please don't believe this because you read it here, rather go do a few searches on your own or read a few books on evolution from authors with opposing evolution views to determine the truth about "species" designation.

Now, if you have not read *The Science Of God Volume 3 - Day Five and Day Six - The Creatures - Revolution or Evolution*, you won't have heard the full discussion of species designation. The trap that many Creation supporters get themselves caught in is the *definition-trap*, and they do this because they are not standing strong on the text that they profess to believe. Part of this problem stems from the late post-Reformation Bibles that have so terribly mangled the Creation text. It is so badly twisted and altered that it appears as if it is a deliberate insertion of lies so as to turn people away by having them believe "fairytales" and by being mocked for their foolish six-twenty-four-hour-day Creation beliefs.

Because those late post-Reformation Bibles have so deeply penetrated the hearts and minds of otherwise very good people, it places those adherents immediately in contradiction, almost forcing them to concede to the species definition set forth by evolutionary "science". When in debate, evolutionists will keep pushing the issue of "How can you get over a million species on a boat that size, clearly you would need more space?" But, as was clearly demonstrated earlier, not *all* animals are the size of a brontosaurus. The dinosaur issue is one of the primary sticking points in the evolution-versus-Creation debate when it comes to the Ark, and it is this area of the topic that can make or break full-scope evolution.

Here is the Bible text regarding the animals to be brought aboard the ark:

"And of every living creature of all flesh, thou shalt bring two of a sort into the ark that they may live with thee: of the male sex, and the female. Of fowls according to their kind and of beasts in their kind, and of everything that creeps on the earth according to its kind; two of every sort shall go in with thee, that they may live."

Creation supporters read "And of every living creature of all flesh, thou shalt bring two of a sort into the ark that they may live with thee" which is taken as an all-inclusive statement, and then the evolution supporters use that all-inclusive interpretation to bludgeon the Creation supporters during debate. The bludgeoning is done through evolution-based species designations. This is where the artistic license for evolutionary species designation pulls a bit of slight-of-tongue and corners the Creation debater in their own words; at which point the obvious arguments made throughout this chapter are asserted by the Creation supporter, thus forcing the entire three-quarter-million species onto the boat with two of each, a male and a female.

You have got to hand it to the Creation supporters because they do a very good job of making evolution's creature body-count legitimately fit in the Ark as per the orders of the evolutionists. And in truth, they accomplish the task in fairness, but the cargo is a little tight. To compensate for this, they will shrink the monstrous creatures down in size by only bringing young dinosaurs and only the young of any other oversized creatures. This is completely fair and is a brilliant solution! This shrinks the 130-foot Argentinosaurus couple down to maybe 10 feet each for a set of adolescent Argentinosaurus, and maybe smaller if they are recent hatchlings.

As you can see, when Creationists chip away at the evolutionary creature descent, it starts to befuddle the evolutionist's arguments. At this point, the evolutionists tend to gravitate to the "oh yeah, well then, why aren't there any dinosaurs today, and how did Noah get them on the Ark since they have been extinct for over a hundred million years and you claim the flood happened only about forty-five-hundred years ago?" This typically leaves the Creationist at a loss for explanation, not because of the vast time discrepancy forced on them by the evolutionist, but rather because they are not able to explain what happened to the dinosaurs.

Where the Creation supporters go wrong regarding this entire cargo-space problem is in *definitions*. While the Creationists do a great job with the animal logistics, there is truly no need to even try to force anything. This all goes back to the issue of who gets to decide what a "species" is? If you have read *The Science of God Volume 3 - Day Five and Day Six - The Creatures - Revolution or Evolution*, you will be well aware of the limited listing of "kinds" used in the Bible.

In the old Latin Bibles, instead of the term "kind", it uses the term "species". Thus, the term "species" was hijacked by evolutionists from the Latin Bible. This means that from a Biblical perspective, the term "kind" and the term "species" are absolutely identical in meaning and value. This places the evolutionists in the position of having changed definitions to suit their own agenda since The Latin Bible long outdates modern science.

Let's take a look again what the Bible says about the creatures that entered the Ark in Genesis Six. "And of every living creature of all flesh, thou shalt bring two of a sort into the ark that they may live with thee: of the male sex, and the female. Of fowls according to their kind and of beasts in their kind, and of everything that creeps on the earth according to its kind; two of every sort shall go in with thee, that they may live."

And in Genesis Seven it says, "They and every beast according to its kind, and all the cattle in their kind, and everything that moves upon the earth according to its kind, and every fowl according to its kind, all birds, and all that fly, Went in to Noe into the ark, two and two of all flesh, wherein was the breath of life. And they that went in, went in male and female of all flesh, as God had commanded him: and the Lord shut him in on the outside."

And in Genesis One it also says, "And God said: Let the earth bring forth the living creature in its kind, cattle and creeping things, and beasts of the earth, according to their kinds. And it was so done. And God made the beasts of the earth according to their kinds, and cattle, and everything that creeps on the earth after its kind. And God saw that it was good."

Now at this point we challenge anyone brave enough to count the "kinds", or in Latin the "species", and prove to the world exactly how many "kinds/species" were brought onto the Ark.

We often get hung up on terms like "all" or "every" and assume it pertains to our time. If you were paying close enough attention to the key "kinds" or "species", you will find that the species quantities are locked up in the evolutionists' forced definition of "species". And, more importantly, the key "kinds" are locked up in the scope of those definitions as shown in the partial animal list at the beginning of this section.

The only "kinds" or "species" listed in Genesis are very broad, let's take a closer look:

- of every living creature of all flesh
- thou shalt bring two, the male sex, and the female
1. of beasts in their kind
2. of everything that creeps on the earth according to its kind
3. and all the cattle in their kind
4. and everything that moves upon the earth according to its kind
5. and every fowl according to its kind
6. all birds, and all that fly
- wherein was the breath of life

As you can see in the Bible's very limited list, the "kind" designation does not list anywhere near a million "kinds", or even seven-hundred-fifty-thousand for that matter.

At this point, it is expected that the first bulleted point will be brought up where it says "of every living creature of <u>all flesh</u>" thus trying to force every evolutionary species to be included in the Ark's cargo. But again, the answer is largely tied up in the list at the beginning of this section. If you scrutinize the evolution species list you might just find out that there are some overlapping species with dual citizenship—or we could call it dual speciezenship in this case. But that's actually relatively low in comparison with the larger number of species that evolutionists have invented. Let's take another look at the list from the beginning of this section, and view it from another angle of species designations of the following: Bears, Horses, Cats, Dogs, Gophers, Muskrats, Rabbits, Skunks, Deer, Monkeys, Gorillas, Apes, Orangutans, Giraffes, etc. Now, if we want to be consistent about terminology, then add 200 Horse species and 339 dog species totaling 539 species of horses and dogs combined and 351 to 564 of all the others listed ranges between 890 to 1130 total species in that list at the time this book was published and depending on how you search the species topic. It is quite possible that if you look long enough you could arrive at figures far outside of those ranges.

Since the evolution supporters are typically trying to force their species designation onto the Creation supporters, it is fair then to take the evolution figures and use the lower range assuming that they estimated high with 750,000 species. Based on the species range of 890 to 1130 species found doing simple samplings in species searches, we find a 22 percent margin of error. By using only 78 percent of the 750,000 estimated species, it leaves only 585,000 that will need enter the Ark, thus giving each creature 28 percent more cargo space 22% / 78% = 28.2%. The big lesson here is that *definitions* matter, and when someone else dupes you into accepting *their* definitions, it can lead you down a misguided path to a great amount of scientific misinformation.

Getting back to the animal list: Bears, Horses, Cats, Dogs, Gophers, Muskrats, Rabbits, Skunks, Deer, Monkeys, Gorillas, Apes, Orangutans, Giraffes with a total species count ranging from 890 to 1130 comes to only the fourteen actual broader kinds in the list that the 890 to 1130 species are derived from. Thus, Biblically speaking, it is possible that of the 1130 alleged species, only 14 kinds where actually on the Ark. And beyond that, by using the same sort of use-my-rules logic, we could whittle the above list down to as little as two or three species if using only the "kinds" actually textually stated in the Bible.

The Genesis Creation account and the Genesis flood account only listed a total of six very broad creature groups. This means that there could be as little as twelve animals that came to the ark, plus of course the others that Noah was supposed to take multiples of.

In no way is this suggesting that the list was limited to only the six broad creature types listed in Genesis, but the probability that it was 750,000 or for that matter 585,000 species, is about zero. The reality is that the "kinds" are more than likely grouped along the lines of Bears, Horses, Cats, Dogs, Gophers, Muskrats, Rabbits, Skunks, Deer, Monkeys, Gorillas, Apes, Orangutans, and Giraffes etc. And we may never know unless God hands someone a manifest from the trip which may not ever have been written down.

Even though this completely removes the need for 585,000 or the 750,000 species that evolutionists dupe unsuspecting Creationist with, we still have to explain the sudden disappearance of dinosaurs.

Now, since "giants" were on the Earth and were considered a sort of mutation of ungodly crossbreeding resulting from the sons of God with the daughters of man, and other such perversions of kinds, we can also consider that what we today call "dinosaurs", may have been similar perversions of creature form, thus they would not have been allowed on the Ark because that was the sort of thing that God was to eradicate with the flood. If this were the case, then getting all of the animals comfortably in the Ark just got a whole lot easier. This contradicts the mention of a dragon-like creature in the book of Job, however, that creature, the Leviathan, was a water-creature that did not need to be on the Ark. There is a great amount of information missing for us to jump to conclusive assumptions. But the reality is, even with the excessive figures that are forced into the debate by evolutionists, the creatures could still all fit on a well-organized Ark.

Chapter 12

A Continental Shift

Jumping back to the effects of the flood, we really need to address some of the problems not yet addressed. It is quite obvious that some sort of continental movement occurred around the globe since the time that Earth came to be. So, the question is not *if* the continents moved, but rather *when* and *how fast* they moved. When getting particular on certain details, it is critically important to get to the root of certain words. Just as with the Genesis Creation text that the late post-Reformation Bibles perverted so terribly (See *The Science Of God Volume 1 - The First Four Days)*, these same versions did the same sort of reinterpretation with the flood text. There is one single word that is key and has the ability to totally change a reader's perspective regarding the flood and how the global geological flood events unfolded, and that word is the Hebrew word "baqa" or Latin's "rupti", both words mean to *break, separate,* or *tear.* "rupti" is connected to the word *rupture* which tends to have us picture something much more violent than *break, separate,* or *tear* might invoke.

In the various versions of the Bible you will see differing language being used that written as per the translators' technical understanding or view of the flood events. Here are some terms used in random Bible translations of "baqa" or "rupti".

- all the fountains of the great abyss were **released**
- all the fountains of the great deep were **broken up**
- all the fountains of the great deep **burst open**
- all the fountains of the great deep **burst forth**
- all the fountains of the great deep were **divided**
- all fountains of the great deep... hath **been opened**
- all the underground waters **erupted**
- all the springs of the great deep **burst forth**
- all the sources of the watery depths **burst open**
- all springs of the great depths **exploded**

There is a chasm of difference between "**divided**" and "**exploded**". If you examine "baqa" or Latin's "rupti", violence is not implied in the text, though it may have been violent. However, using terms like "**exploded**", immediately forces the reader to assume extremely violent activity. The potential pressures that could be generated, as discussed in Chapter 4, are maximum figures and are calculated as if the supporting rock matrix instantly crumbled, and if that was so, then violent it would be. However, it is possible that the water only relatively gently bubbled up from any fractures in the land-masses. We also have to consider that since the water had to come from some logical place, then beneath the tectonic plates is about the only viable option remaining. Since the continents are miles thick, the ejected water would have to make its way through two to three miles of ocean depth before it made it to the surface if it was forced out from under the tectonic plates at the edges of any existing continent(s) at that time. This means that a good portion of the flood waters may have had relatively little disturbance on the surface. Think of this like filling a half-full bathtub with additional water using a garden hose placed under the water, shooting horizontally across the bottom of the tub. The higher the water level, then the less disturbance the water from the hose will create. This somewhat contradicts what is mentioned in chapter four implying that the "fountains of the great deep burst forth".

The term "fountains" also invokes pictures of shooting water, but not necessarily as violent as does the term "exploded". Latin "fontes" or Hebrew "ma yan" are the terms used for "fountains", and neither word demands a violent release, or for that matter a gentle release, though they are both far gentler in definition than is "exploded".

The reality is that there was likely a vast array of effects ranging from barely noticeable on the surface, to extremely violent. However, if some sort of super continent did actually fracture and break apart, then the water would come from that fracture crack. But, once the crack was filled with water then the pressure dynamics are still the same as if the water is ejected at the ocean floor level. When we allow nature to be nature and physics to be physics, it all starts to make logical sense. But when we take an all-or-none approach, then we are forced to invent wildly unlikely conditions.

What Do the Ocean Maps Say?

When examining maps of the ocean floor, it is clear that the current continents have moved over time. The two most important questions to ask are: What amount of time did it take for the continents to be positioned as they now are? And, what was it that actually moved?

The ocean-floor maps present clear evidence that things did move in the past, and possibly still are moving to this day. We typically view this as the continents are "drifting", but "subduction" theory implies that the continents are being both pushed and carried, and in some places, it is believed to be a sort of collision of direction. It is a complex task to attempt to interpret the past movement of the Earth's surfaces both above and below water because you cannot see it all at once on a typical globe. And using a flat map is very deceiving because flat maps either stretch the map image progressively further as you approach the poles, or the flat map breaks up the pole areas into segments, both of which make it hard to determine motion direction and what exactly might have moved and in which order that movement occurred. Ocean maps show trails of movement thousands of miles long, many of which appear illogical. However, when such irrefutable naturally occurring and massive amounts of evidence appear to us to be illogical, then it's time for us to reevaluate our own theories and perspectives. Things are not always as we choose to interpret them, and that has never been more obvious than when reading the various theories about the position of the continents and studying the ocean floor. Sometimes things in reality just don't match our theories the way we would like them to. When we finally see the light then reality makes a whole lot more sense.

There are so many factors in trying to calculate the topography of the depths of the oceans and land elevations that getting the true accurate full picture is unlikely to happen anytime soon. We have to take into consideration that the Earth is a ball of mush floating in space. If you could set our Earth on a giant table with the gravity pull equal to Earth's, then the Earth would no longer be a ball because the gravity would deform or flatten it considerably.

Even when calculating the effects of the polar ice caps melting, we have to realize that the Antarctic continent will rise somewhat to compensate for the lost pressure that the ice was exerting on the land. This is not a small area, so the overall effect can be substantial, thus if the ice on the poles did ever completely melt it might not raise the ocean level much at all, or it could make the ocean level rise more than what the ice melting alone would do. It really all depends upon how you calculate the effects. However, while these calculations can be *mathematically* done, that is insufficient when we do not know for sure how the adjacent land will react to the reduced polar pressures. As the polar pressure is reduced, not only will the continent raise slightly, but so will the surrounding ocean floor.

Then as the ocean level rises, it will place more pressure on the overall ocean floor, causing other areas to compensate until it all comes into gravitational balance once again. It is impossible to accurately estimate the effects of such major changes on the surface of the Earth. Our best-guess estimates, while taking into account as many factors as we are aware are needed, is the best that we can do, but it fails us in reality.

Shrinking and Sinking

Regarding the "great deep" mentioned in the Bible, we need to grasp the height of the supporting structure that would have collapsed beneath the tectonic plates, which could easily have been well over a thousand feet. If you like math problems, then here are some rough figures for you to run some calculations with:

- Diameter of Earth 7,918 miles
- Circumference of Earth 24,901 miles
- Surface of Earth 196,900,000 sq miles
- Water on Earth 139,382,879 sq. miles about 70%
- Land on Earth 57,517,121 sq. miles about 30%
- The average depth of the ocean is: 3,684meters or 2.26 miles
- Average land elevation above sea level is 847 meters or 2,778 feet
- 3107.64 cu miles of water per inch over entire Earth surface
- 1,000 feet of water is 12,000 inches
- Polar caps melting raises sea level about 253 feet
- Antarctica is estimated to have 7.2 million cubic miles of ice calculated from 2.4 million square miles at about 3 miles thick
- Calculating the **ocean bottom** and dry land surface as a single level surface, the average height is 2382.3 meters or 7815.9 feet BELOW sea level
- Sinkholes in our modern times that are naturally occurring indicate to us that the Earth's crust does display some level of porosity. Lava rock is also very porous and lightweight.

Figure 6. First table of numbers for your mathematical enjoyment!

Regarding the beginning of this chapter discussing the various terms used to describe the "fountains of the great deep" having been "broken up", we need to be fair in analyzing the Bible's text, so we have to look closely at these few words: "all the fountains of the great deep were broken up". Since it says "all the fountains" we will make the assumption regarding the flood, that this was widespread around the globe wherever the matrix of supporting rock existed. Since the text says that "the fountains of the great deep were broken up" we will attribute the being "broken up" to "the great deep". This is consistent with the logical source of the water and what would need to occur in order for that water to raise sea level relative to the dry land at that time.

In Noah's day, the land likely had a relatively flat surface in contrast to today's miles high mountains. Also, as the land dropped lower from the rock matrix crumbling beneath the land masses, it would logically cause a dual effect where when the land protruding above the water settles, it would logically raise the water level at the same time. So as the land is sinking the water is also rising. This effect is equal in volume and it makes a significant mathematical difference regarding having enough water to cover everything by at least fifteen cubits.

When it comes to the flood, there is simply no knowing the exact topography during Noah's life before the flood, but in understanding that the movement activity on Earth caused today's mountains to form, regardless of when, we can safely speculate that the mountains did not exist then as they do now. Many people when analyzing this will either move the continents over millions of years via "continental drift", or they do it all at the time of the flood. The movement evidence on Earth was likely left behind somewhere in between now and when the flood began, but probably much closer to the flood and it probably took a while.

Divided Migration-Phaleg

In Genesis Chapter Ten, we have this peculiar statement "And to Heber were born two sons: the name of the one was Phaleg, because in his days the earth was divided..." The name "Peleg" or "Phaleg" when translated is replaced with *divide*, *earthquake*, and *stream*. This would fit with the major continental fractures seen in our modern era. "Phaleg" was born about one hundred years after the flood. This suggests that there was longer-term geological activity than just during the year of the flood. The topography was probably changing on an ongoing basis until everything finally settled into place. This means that geological activity could have been fairly active for many hundreds of years following the flood and after Noah and family exited the Ark. If in Phaleg's time continents did begin to "drift", it would explain how people ended up on very distant land masses where they started their own civilizations. This is not implying that it was like folks departing on a ship, but rather that as the land masses slowly moved apart, people could have boated across to-and-fro from land mass to land mass until it became impractical to do so due to the expanding nautical distance. This comfortably explains migration capabilities to currently distant places like Australia, without them getting lost at sea.

Chapter 13

Oceans of Evidence

The Ocean floor has a great deal of evidence in the form of movement marks and sediment, and the land has layers of geological strata that have entombed mountains of evidence, and the Bible has an account of alleged Creation and an account of an alleged worldwide flood. How we choose to interpret all of that evidence will limit our conclusions by what we personally are willing to allow in as evidence, by how we interpret that evidence, and by the assumptions we make due to our prejudicial views.

There's a good variety of explanations for all of the creatures caught in the layers of sediment, ranging from billions of years of evolution, all the way down to getting trapped in the sediment and having it all occur due to the Biblical flood in a comparatively short period of time.

Caught in a Rant

When we listen to evolution-versus-Creation debates, it is common to get caught up in someone's impassioned rant. But let us not allow either the religion of evolution, or the religion of six-twenty-four-hour-day Creation, to cloud our better judgment. Someone's passion and strength of belief has no bearing on the actual truth, regardless of whether or not they are correct. If you're right, then you're right, and if you're wrong, then you're wrong no matter how badly you might wish otherwise. Belligerence, shouting, mocking, and lying by fudging the facts will not make you right when you are wrong. For instance, evolutionary naturalists are very disingenuous when they refuse to apply the same standards to their own theories that they force onto Creationists. People must be called to task on all of their inconsistencies when they state their theory as "fact". Truth will always win in the end, and the sooner we accept the truth of using fair logic, then the sooner we can move on in our understanding of the actual origins of all things.

Science in Flux

There have been many discussions and debates over the years where people on the evolution side claim scientific superiority and promote most of their thoughts and theories as "settled fact". But when you ask them if things ever change in science, they have to admit that things do change and have changed. This is a self-made trap for evolution supporters, because if something changes, then it was not correct before, thus it had been wrong in the past and was, in fact, not "settled" or "fact". So, what is to stop the newly adjusted theory from also being found to be incorrect in that same way at some point in the future?

Covering the Hills

The geological layers are probably our best evidence for either Creation, or for full-scope evolution. This is the death nail for either side depending upon their ability to prove either the short-term deposition of layers, or a long-term deposition of layers. If the layers can be irrefutably proven to have been long ages between each layer, and even long ages for the layer to be deposited, then Creation and the flood are doomed and are lies. But if short-term deposition can be proven, then the entire evolution house of cards will quickly collapse with no hope of ever being rebuilt, because the entire evolutionary argument is absolutely, fully, and completely built upon the foundation of long ages that are based upon the sedimentary rock layers that they assume to be millions or even billions of years in the making.

When viewing the geological layers, there are many discrepancies in evolution's long-age deposition theory. But on the other hand, proper Creation and flood theory have only one discrepancy, which is not really a discrepancy, but rather an inability to prove that the layers are not as old as the evolutionists' claim they are. This, you will find, is the single biggest fear of evolutionists. If a Creationist can prove their case regarding the age of the layers, then the debate is over for all time. Prior to radio-carbon-dating the evidence was typically in favor of flood theory, but since the advent of high technology this view has changed.

If we have no perfectly reliable means by which to date stone, then we have to look at logical arguments only. Actual logic is where the evolutionists have many flaws in their theory. Since each of the layers are said to be millions of years old, we need an explanation as to how certain creature fossils still show fairly high detail. If the layers took millions of years to form, then many of the bones and features of the creatures would have decayed completely, or at the very least substantially, before eventually being completely buried and entombed in the rock layers forming the fossils that we find to this day. And if it were the case that the layers took millions of years, then there would be progressive decay and erosion of the fossilized bones as the elevation of the creature bones progressed as the bones were slowly being buried by sediment or settled dirt over many years. When these points are asserted, then the evolutionists will talk about cataclysmic events that suddenly deposited a thick layer of sediment, resulting in the rapid burial of the found creature fossil. But then they are faced with the problem of larger items and creatures piercing several layers. If each layer took millions of years to form or even millions of years between layers and the creature died and was partially buried in the lowest layer that their fossilized bones are partially caught in, then their actual bones would have decayed at a different rate in each previous layer, but this is not what we see today.

There are also petrified trees that pierce multiple layers. If a wood tree was suddenly partially covered, it is logical that the buried part of it would be preserved in that layer, but if the remaining exposed portion of the tree sat for even four or five decades, surely it would have rotted away leaving only the part of the tree's trunk buried by the first layer. If subsequent layers were deposited millions or even only thousands of years later, there are simply no rational explanations that would allow a tree trunk piercing multiple layers to occur, this is especially true regarding the amount of decay from layer to layer.

Then when you take into consideration the nature of the layers, the layers that have strong traits and are very widespread within any given region are often rightly attributed to catastrophic flooding as stated by most evolutionists. However, in that case, it is always claimed to have been a large but regional flood. The idea of regional flooding does not preclude a global flood in that regard.

If a global flood did occur at any point, it does **not** mean when Noah left the Ark that everything was done and settled into place at that point, nor does it mean that regional flooding didn't occur as the flood was underway the first week or two.

As was indicated in the previous chapter regarding "Phaleg" being named after the division of land, the flood likely caused centuries of periodic and fairly violent geological activity following the forty days of flooding. It is critical to understand that this is a big Earth, and when the flood waters began to recede and the continents were shifting and settling into place there would have been substantial activity that was far worse than any of the worst earthquakes we have experienced in recent times.

Even an extremely violent earthquake will do little harm if you are living in a simple structure. A severe quake might physically knock you off your feet, but when it is done then you get right back up and fix your simple home that was made of stone, grass, twigs, or branches. And since all of your materials are close at hand in the form of your collapsed home, there will be very little work required in order to restore your home, and that's if you are not living in a basic cave. Today, however, things are very different because we have tall buildings and highways and complex built homes that get damaged in earthquakes, which greatly inconveniences us. This is something that would not have been the case in the early years after the flood because our modern-type complexities could not yet have been built immediately after the flood.

As the continents were settling, the waters simply would have not been able to recede much at all from evaporation alone. So at some point we need to find someplace for that water to have gone. But let's ignore that for now and consider what would happen as the waters receded. As many people are aware, if there is an earthquake near the coastline, it will often produce varying sized tsunamis depending upon the strength of the earthquake. Tsunamis can travel inland as far as the volume of water, speed of the wave, height of the wave, and topography will allow. A strong tsunami can carry debris and silt with it and even pick up and push more silt and dirt as it travels, leaving thick layers of sediment with all sorts of artifacts within the newly deposited sediment.

There is a clouding of thought that occurs when we analyze the time surrounding the flood. When people picture the entire flood scenario, the mental imagery is typically divided into three basic groups: before the flood, the actual flood year, and after the flood. But seldom do we focus on the actual forty days during the flooding.

Thinking through those forty days is critically important. If you fail to grasp the issues that might arise with plate tectonics, water dynamics, chemistry, and the creatures' ability to survive, you will certainly arrive at wrong conclusions. This is a tough topic to tackle in our modern era because there is a very strong inclination to assume that "science" has things all figured out. And since full-scope evolution is the modern foe of Biblical Creation, Biblically speaking we tend to connect the flood topic to the Creation topic, and then because of our modern-era view regarding evolution, these two areas of study, the flood area of study and the evolution/Creation area of study, end up being subconsciously inseparably connected in our minds. But in Truth they are actually not connected at all.

From a Biblical standpoint, there is no connection whatsoever between Creation and the flood. However, when discussing evolution, the evolutionary evidence is based largely upon what is found in the various layers of sedimentary rock. This is why if the rock layers are ever proven to have been all deposited in a relatively short period of time following the flood, then the long-ages that full-scope evolution requires would not fit with the actual short-term timing model. In other words, if the sediment layers that all of the fossils are trapped in can be proven to have occurred within only a handful of hundreds of years, then the entire evolution industry has no basis for their required time estimates, and because of the loss of that timespan, their entire theory is dead. This is true even if the layers were deposited, let's say, over a period of a couple of thousand years. And even if that couple of thousand years occurred millions of years ago, it still devastates *all* of the evolutionary arguments. For full-scope evolution to be true, the layers require long ages between them. Without the long ages between or during layers, evolution, as proposed, simply could not occur.

If you use logic and basic common sense when examining the layers and the fossils, it is very clear that each layer was deposited quickly. One way we know this is through the fact that some of these layers are chemically bound in the way cement powder is. Cement is not something that man invented, cement is something that man discovered and modified to suit our needs. When certain materials are heated and then later wetted, they bind together and become rock, such as limestone.

When there is a limestone layer with creatures caught in it, those creatures become encased in the newly formed stone. And as with cement, it takes time to solidify. When we manufacture cement powder, it is treated with other chemicals and heat treated and then kept dry until it is needed. Modern cement is designed to cure and harden very quickly, in only a few hours or less. But when you pour a cement floor, depending upon the air temperature, the cement can stay in a putty state for days when it is very cold when no additional accelerators are added. Naturally occurring cement-type material, however, will not harden so quickly and will stay in the putty state much longer, especially when cold. When heat reactive material is deposited it could sit for weeks in varying degrees of being in a putty state until it is finally hardened, and that final hardening can take years. Sediment in a putty state allows for impressions to be made that are located both below the creature and then above the creature but on the underside of the newly deposited sediment layer, which would obviously conform to the structure of the dead creature that the layer settles upon.

When traveling abroad and paying close attention to roadway cutouts in the hills, you will often see several thick layers of rock.

But within those thick layers there are many relatively thin layers with vegetation and insects and small creatures and fish fossils often being found between those thin layers. When observing roadway cutouts, we often see thirty or forty-foot-high walls of rock on either side of the roadway, but that fails in comparison to areas such as the Grand Canyon or the remaining platforms you can witness in the Arizona desert. These layers are numerous and widespread. And on the currently standing landforms, the amount and manner of erosion is not consistent with the super long-age erosion patterns that would have had to have occurred using evolution's required long-ages.

Figure 7. Highway cut-away in Iowa showing flat, clean layers.

If the layers of sediment occurred millions or even as little as thousands of years apart, then there would be a great deal of erosion between layers, yet what we see is little or no erosion through nearly all layers.

Then we have the seemingly unexplainable landform pillars with huge rocks precariously balancing on top of the pillars. Following is a photo taken of what happens when rain falls upon stones and the dirt around them washes away. This familiar occurrence is what is often seen in deserts but on a much larger scale in those desert situations. Do a search for rocks balancing on pillars in the desert and you will see a striking resemblance to the following photo. The effects seen in the Figure 8 photo occurred during a single rain storm.

Figure 8. Balancing rocks after a single rainstorm.

What we see in the erosion patterns remaining to this day, is that the layers had to be deposited quickly and would have had to have been in varying degrees of putty state in order for the landforms to occur that we now see in the Arizona desert and other such places. The flood would not have caused equal distribution of layers of sediment all around the world. Any layers deposited would be consistent with the events and materials surrounding any one area. Keep in mind that as the supporting rock matrix was collapsing, it would have created a great deal of material from the underlying matrix structure that would have turned into silt, stones, and rocks on its way out from below the tectonic plates. That silt is some of the material that we see in the layers today.

What everyone needs to understand is that if the flood actually occurred due to the reasons stated in the Bible along with its complimentary ancient writings, then we will eventually find evidence of such. And in fact, we do find evidence that would indicate that the tales of the pre-flood era are at least partly true. Following in Figure 9 is a photo of a footprint found in granite rock. No credit source for this photo could be found.

This print is on the side of the rock and the granite is mushed forward in front of the toes' impression much the way you will see when you walk in mud. This of course is not a big deal except for a few key points: The first point is that the footprint is in granite rock, this means that at some point this granite was soft enough for the impression to occur. It may have burnt the feet of the fleeing owner of the foot that made the print, but the footprint is there in the granite for all to see and judge for themselves.

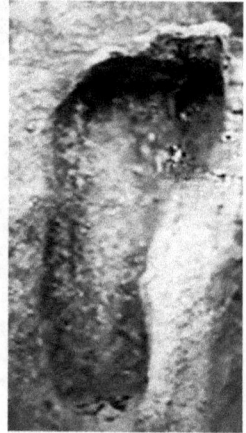
Figure 9. Empuluzi, South African footprint.

The second point is that the impression is currently vertical and it is unlikely that a person laid down to make the print, especially when considering the way the material mushed out in front of the toes' impression. This means that when the material was lying flat and was stepped on, it was soft enough to make a footprint, and then it subsequently solidified at some point and the newly solidified rocky material was later forced out of its original horizontal position into its current vertical position. Yet there are those who will deny that this is an actual footprint and claim that it is erosion or some other nonsensical imaginative non-footprint explanation. You have to imagine pretty hard that this is not a footprint to make yourself believe that it is not, especially when you consider that there is another partial opposite footprint of the same size at what would be a considered only a step away.

And finally, and most importantly, this footprint is over 48 inches long (1.2 meters). Yes, the footprint in the photo is *actually over four feet long*. Could someone have carved this foot print to dupe the world? Possibly, but highly unlikely. If there ever was evidence of pre-flood giants, this would be it. And since there is another partial similar sized footprint of the opposite foot at what would be a giant's step away, it is fairly compelling evidence of the Biblical pre-flood giants roaming the Earth.

But, you be the judge and search for the "Empuluzi, South African giant footprint" and see for yourself if you think it is carved or had eroded, or if, maybe, it is a real giant's footprint that is possibly from the pre-flood era.

Chapter 14

Life on Board

How did Noah and family survive for an entire year without any food on board? They didn't have to worry about food because their food was living all around them. God instructed Noah to bring extra animals for that very purpose, and living animals don't spoil. Noah was also told to bring other edibles.

A Year on Board

A year on board the Ark is a long time to be confined, but the entire family would have been busy caring for all of the many animals. You can imagine, towards the end of the year afloat that they might have been getting anxious to set foot back on land. If the Creation supporters foolishly try to force the evolution supporters' species count aboard the Ark as calculated in earlier chapters, then Noah can probably get the job done, but it would be very uncomfortable. And since we have used the entire cargo area for animals to inhabit, we have one looming problem, where is the animals' food stored?

Since most of the animals are vegetation-consuming creatures, where did they get their food from? We have to take into consideration that to feed that many creatures would have taken an enormous amount food. As mentioned in an earlier chapter, it is highly unlikely that the Ark had two of each of the evolutionists' definition of "species" on board. The amount of kinds technically would not have to be any more than the six kinds mentioned in the Genesis Creation and flood accounts. This would have left an enormous amount of space for food. The reality is that if the flood did occur as stated in the Bible, then the "kinds" would have been more broadly specified in the form of dog or cat or horse etc. rather than the various modern species definitions of each "species" of cat. Reducing the species down to reasonable divisions with the obvious child-observable-kinds is more likely much closer to the actual cargo manifest. This would allow for a proper balance of adequate space for the creatures and plenty of room for grains and hay etc. While the Ark was likely a bit crowded, being reasonable with our unfounded mental distinctions of the "species" solves all cargo logistics problems.

What's for Dinner

Placing enough food on the Ark to last forty days is not a problem, it is the other three-hundred plus days that they were afloat that presents a problem for Noah and family and their precious cargo. What food options did they have? Since they knew what was coming, it is very possible that Noah family and fattened themselves up for survival purposes; this way they could consume far less than their daily calories needs would normal require and still thrive for many months on very little food. But we need not go even to that length for their survival. Possibly having the bottom deck open to the free flow of sea water, they could have had an area that would allow them to access sea creatures, like fish. Thus, Noah and family could have had abundant access to seafood as well as using the animals that they brought on board for food, and they also likely had some vegetation and grains stored for themselves.

But what about the carnivorous animals, what did they eat? Most creatures will eat what they can if desperation ensues. But let's not even go there. Since many animals such as rabbits can multiply very quickly, there could have been an abundance of creatures on the boat in a matter of months. And in addition to that, since God had the creatures come to Noah, it is possible that many were already pregnant and could have given birth as the rains began. Reproductive rates on many small animals are so short that within a few months there could have been thousands of additional creatures born, and taking that to second and third generations of those fast-gestating creatures, Noah and family then had many thousands of additional passengers that could have been eaten by the carnivores. We also have to realize that they could have initially fed the carnivore's some meat from the animals brought along for food. And then we also have potential seafood that might have been caught and used as food for the carnivores.

Many land, air, and water creatures have speedy reproduction rates and quickly come to reproductive maturity depending upon the reproductive cycle time. For instance, they could have had several chicken reproduction cycles from one chicken lineage alone, but now consider that chickens lay eggs every day or two and in a month you have twenty to thirty eggs and maybe more all gestating. So, in less than six months, all twenty to thirty of the eggs could have hatched and the chicks come to maturity and then go on to lay eggs of their own. As you can see, there was no shortage of food on the Ark.

Gardens Afloat

Since Noah and family had no ground to till in order to plant crops in, they had to rely on the foods they brought with them plus any seafood they could catch, along with the meats, eggs, and milk being created as mentioned in the last section due to rapid animal reproduction. But oh how they would have longed for fresh vegetables!

Since the Ark was unlikely to have had any sort of an outdoor deck around the living quarters, as it often depicted in children's books, there would have been nowhere on the top deck to plant even the smallest outdoor garden. And since the text seems to have very specifically only instructed Noah to make a single window in the Ark, it is likely that there was no sunlight available inside the Ark to shine on any garden that may have been planted inside the Ark.

But let's stop for a moment and consider a few things. First, while the Ark was huge, it says only one window and one door were included in the Ark design, there could have been more, but that is not indicated in the text. Since it was probably very dark in the Ark being only lighted with a flame of some sort and only when the family cared for the animals, the animals probably got a lot of extra sleep and some animals likely hibernated in such conditions for most of the duration of the voyage. However, the other animals still have to eat. Animals that eat make lots of manure.

Did Noah and family toss the manure over board? Possibly not. Since according to the design specifications for the Ark, the roof would have had a very low pitch, just enough to shed water, and if "the fountains of the great deep" did shoot up at all, then it is possible that silt and dirt and volcanic ash settled on the roof of the Ark. It would not take much silt to suit Noah's needs. And if they were thinking ahead, then they possibly even had strips of wood on the roof to keep a relatively thin layer of dirt on the roof, and thus they may have preloaded the roof with a few inches of dirt themselves.

But let's assume that they did not specifically plan for that ahead of time, then if any silt did get on the roof and they carried the manure from the animals onto the roof, then within a matter of days they could have had ground fertile enough to begin to grow food. They could have even covered the roof with pitch so it would not leak and then covered it with dirt and had grass growing on it before the flood even occurred. Imaginative? Maybe. But very plausible and easily doable if needed.

Based upon the figures used to calculate cargo space and payload capacity desired, we used dimensions of 525 feet long by 87.5 feet wide. That's 45,937 square feet of garden space or just over an acre. That is a huge garden by any modern measure. Noah and family had ample food stuffs already and potentially had no shortage of fresh vegetables. In fact, if the scenarios being proposed here were known and understood by the entire Noah family, they could have had starter plants just waiting to be transplanted into their roof garden when the rain subsided, thus cutting weeks off of the wait for freshly grown vegetables.

If starter plants were planned and prepared ahead of time, then something like tomatoes, peas, and radishes etc. could produce fresh food in only about a month after being transplanted. Besides the obvious logical and likely fact that they brought a lot of food along with them, they could have had started the plants before the flood, but more importantly, they also could have started plants while it was raining and then could have easily transplanted those newly started plant onto the roof after the rain subsided. Plants do not need sunlight to for a couple of weeks in order to thrive until they have sprouted well above ground. If they planted starter-plants the day the flood began, they could have transplanted them and had some vegetables within a month after the rains stopped. If they planted starter-plants before the flood began, with some of the slower growing items they then could have begun eating those items about the same time as the fast-growing items.

Then we also have to consider that some edible fungi do not need any sunlight to thrive, and thus they also could have used fungi as a food.

With all of the potential options for food, it is quite apparent that Noah and family and all of the creatures could easily be well-fed with plenty of quality food for the entire duration of their world cruise. They would have been well-fed and had no problem staying alive and would have been very active with new life burgeoning all around them during the entire voyage.

Now let us take additional consideration of modern-day high-density gardening techniques, often referred to as "high-density urban gardens" that can produce *substantial vegetation in every square foot* of garden—Noah and family were not starving.

All of what is mentioned in this and other chapters are not intended to have you believe that this is exactly how things happened, nor is some of it written in the Bible. Rather, the ideas being proposed here are viable very possible options that could easily have been used by Noah and family if needed.

Chapter 15

Steady as She Goes

When thinking about the Ark adrift on an endless ocean for an entire year, surely they would have encountered some rough seas. Could this glorified raft stay afloat?

As speculated in an earlier chapter, the Ark was not likely to be a hollow-hulled vessel. Instead the entire structure was likely based upon the buoyancy of the wood used in the form of a thick barge/raft structure made of solid light-weight wood covered in pitch. Regarding the buoyancy, we have two additional factors to consider. In standard clean water the buoyancy figures previously mentioned apply, but the buoyancy amount could have increased substantially if the water was salty and/or had a lot of suspended silt in it. Salt water alone could increase the buoyancy roughly two percent, so depending upon the cubit size used, it could easily have added up to 800,000 pounds of payload capacity to the Ark's payload depending, of course, upon the wood density and actual size of the cubit that they used—that's over 400 cows worth of extra animals, **MOOO**! The Ark could have been riding quite high in the more turbulent days of the flood until the suspended silt settled out of the water. But all of that is a separate matter than the Ark's actual stability.

Could an Ark built upon a barge/raft base as is suggested in this book be nearly impossible to capsize? Or would the Ark overturn at the first waves it encountered.

This is very easy to test and would be an excellent teaching tool for students of any age whether you are for or opposed to the Bible's flood account. Simply use a 6-inch-wide by 1-inch-thick piece of typical softwood pine lumber from your local building supply store. The lighter the wood, then the more load your Ark model will carry. Then cut your wood to about 33 inches in length. Your finished Ark-raft base will be about ¾ inches thick and about 5 ½ inches wide and 33 inches long. This is the rough size of the Ark at roughly a 1:191 scale of a 525-foot Ark (160 meters). Then cut two pieces of the unused leftover board each about 8 inches long and firmly attach one on top of each end of the 33-inch-long board. Now take some plastic poster-board and construct a 33-inch-long by 5 ½ inch wide frame that is 2 inches high and attach it to the wooden base. If you want to be more particular about it you can cut the poster-board tray ¾ inches taller and notch out where the negative ballast is on each end. See the following Figure 10 illustration.

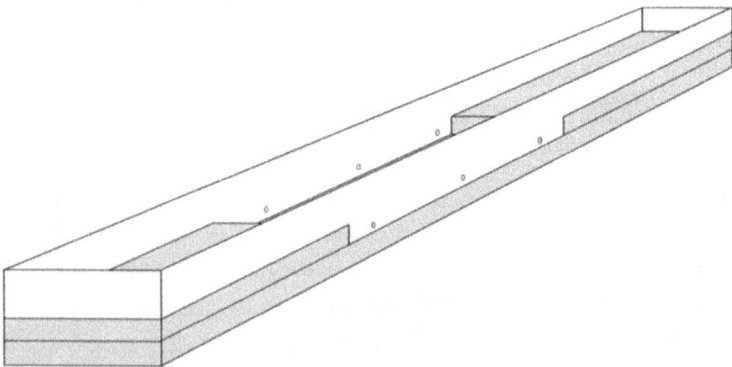

Figure 10. Ark buoyancy test using pine base with plastic poster-board sides. Notice the water holes in the amphibious creature pool basin for the free-flow of water. This ark design deliberately allows water inside—It is unsinkable!

Then securely attach the plastic poster-board frame to the raft base with staples or hot glue and add a flat roof. But make sure to add a few heavy coats of paint inside and out to emulate the pitch that was instructed to be used. If you fail to do this, then your Ark's base will quickly absorb water and lose a fairly high percentage of its buoyancy.

To repeat: It is important to attach the shell well so that it doesn't rip off when you toss it into the water, as you test its sea worthiness. As well as making sure to punch holes in the sides of the pool area so that water can freeflow into and out of the Ark.

Now, once your paint has dried, place your Ark in water and load it with weight, while balancing the load side to side and front to back. Do this until the pool area is near full to its top with water. It is important that the water does *not* flow over the raised floor area around the pool basin, those amphibious critters also need dry space to hang out on. When you have finally achieved your payload limit, then take your payload weight and weigh that, and then multiply it times your model's scale size three times.

If your model is 33 inches long (0.8382 meters) and you are using a 21 inch cubit (0.5334 meters), then your scale is 1:191. If your Ark held a payload of 1.7 pounds, then multiply 1.7 x 191 x 191 x 191 = 11,845,380 pounds of allowable payload. The number can be very different depending upon what type of wood you use. Harder very dense pines will carry far less than something like Blasa wood is able to. So be aware, that if you are not very familiar with wood and end up using a dense hardwood, then your payload will be very low. The wood that Noah used is believed to have been less than half the weight of water for the same size volume of water.

Data Table for Scaling Cubits

Using 10-foot per floor heights, except for first/bottom level which is partially raised so that there can be a shallow water pool in it for amphibians, will produce solid Ark base of a pine wood weighing under half the weight of water; thus, only about 40 percent if the base volume will be under water when no payload is applied. The Ark model shown in photos in this book is made using materials found at most craft stores, such as tongue depressors, Popsicle sticks, paint mixing sticks, and various sized dowels. You will notice in the following figures, that when using a 12-inch cubit, the Ark would essentially have no room remaining for the actual base, thus a 12-inch cubit would not be a viable cubit size when using a 10-foot ceiling height. In that case you would be forced to reduce the height of each floor to make room for the Ark's base-raft.

Some basic figures needed to calculate Ark payload weight and volume

	12" Cubit	18" Cubit	21" Cubit
Model Cubits Translator			
Each log is 5 cubits	60"	90"	105"
Each paint stick is 1.15 cubits	13.8"	20.7"	24.15"
Each Tongue depressor is 0.625 cubits	7.536"	11.304	13.188"
Each Small depressor is 0.55	6.6" (0.55)	9.9" (0.825)	11.55" (0.9625)
Each Popsicle stick is 0.7142	8.5704"	12.8556"	14.9982"
Ark Height	30'	45'	52.5'
Ark Width	50'	75'	87.5'
Ark Length	300'	450'	525'
Ark-Base square footage	15,000 sq. ft.	33,750 sq. ft.	45,937 sq. ft
Head-space with 10-foot floors (3 levels)	9' 5.4'	9' 2.3"	9' 0.45"
Remaining space left for solid base	0'	15'	22' 6"
Available cubic feet of space for solid base	0 cu ft	506,250 cu ft	1,033,593 cu ft
Cubic inches in a US gallon	231 cu inches		
Cubic inches in a one-foot cube 12 x 12 x 12 =	1728 cu inches		
Gallons in a one-foot cube 1728 / 231 =	7.4805Gal.		
Pounds weight US Gallon of sea water	8.34#		
Water weight in a one-foot cube	62.38737#		
Pounds water weight of volume of Ark base	0	31,583,606#	64,483,195#
Solid-base weight of Southern Pine (36.7#ft)	0	18,579,375#	37,932,863#
Gross Payload using Southern Pine	0	13,004,231#	26,550,332#
Gross Payload using round Southern Pine logs no nesting (21.5% empty space where water can fill)	0#	10,208,321#	20,842,010#
Solid-base weight of Northern White Cedar (21.4#ft)	0#	10,833,750#	22,118,890#
Gross Payload using Northern White Cedar	0#	20,749,856#	42,364,305#
Gross Payload using round Northern White Cedar logs no nesting them (21.5% space where water can fill)	0#	14,584,809#	33,255,979#
Scale-Model Cubic feet of floors and ceilings not included in solid base	24,750cu ft	83,531cu ft	132,644 cu ft

Scale-Model Cubic feet of walls not included in solid base. 11,550 cu ft .... 25,987 cu ft ....35,371 cu ft
Total Cubic feet of floors/ceilings and walls........................ 36,300 cu ft .... 109,518 cu ft ....168,015 cu ft
Total wood weight of floors/ceilings and walls ....
Using Southern Pine @ 36.7#ft... 1,332,210#.........4,019,310#.............6,166,150#
Using Northern White Cedar @ 21.4#ft .... 776,820#...........2,343,685#.............3,595,521#

Net Cargo and passenger Payload using Southern Pine
round logs no nesting and deducting floors & walls weight. -1,332,210#.......8,984,921#.............20,384,182#
A 300 Ark foot using Southern Pine floors & walls carry ..... 932,451# (has no base)
Southern pine would be under water on a 300-foot Ark

Net Cargo and passenger Payload using
Northern White Cedar.. -776,820#.........18,406,171#.............38,768,784#
A 300 foot Ark Northern White Cedar floors & walls carry   1,487,841# (has no base)
Northern Cedar would be under water on a 300-foot Ark

On a 450-foot Ark, 10 feet is equal to 6.67 cubits
On a 525-foot Ark, 10 feet is equal to 5.71 cubits
The floor height in the model shown in this book is just over the thickness of a 1/2" dowel.
Figure 11. Second table of numbers for your mathematical enjoyment!

Adrift for a Year

As you will see, if you either picture it clearly in your mind, or if you actually take the time and enjoy doing the experiment as described in the previous section, the Ark is nearly impossible to capsize. Is it possible to place it gently in the water upside down and have it stay? Yes, but the ability of this basic design is far more stable than any conventional boat design. From this simple experiment we can see that capsizing would not be a problem for the Ark unless it got broadsided by a massive cresting wave that towered over the Ark and turned it upside down. You will notice that if you set your Ark on its side in the water it will always end up proper, and even if you set it in the water at slight angle favoring upside down, it will still right itself.

As far as their ability to stay alive and upright in the boat, along with the abundant food discussed in the last chapter and the stability and strength of the Ark design, Noah and family would have had no problem safely surviving adrift for an entire year.

Weathering the Storm

Now, regarding weathering the storm, based upon the Ark design mentioned in this book and the obvious stability mentioned in the previous section, the only real hazard for the Ark and crew would be severe weather with winds causing extreme waves, and possibly very cold temperatures according to some people.

There are many evolutionists who fear the flood being proven true because it would utterly shatter their entire belief system. That fear results in many people making outrageous claims such as "The temperatures would be so cold that they would have frozen to death." Beside some obvious oversights in this, let's address the temperature issue. If the Ark was built using mortise and tenon construction with a heavy beam structure and then additional mortise and tenon walls of about 2 inches thick and sealed with pitch, the Ark would have been very well insulated, and with all of the animals producing body heat they could have easily kept the Ark at reasonable temperatures, so they could easily have survived sub-zero temperatures.

Now, let's examine the reality of the low global temperature theories. With the friction of movement and pressures involved with the "fountains" being "broken up" in the "great deep", it would have heated the water to fairly warm temperatures as it was being ejected from the "great deep". The friction and pressure would produce a lot of heat and when that mixed with any ice that might have existed and the current ocean temperatures produced at that time, it would have been an ocean with temperatures somewhat warmer than the temperatures we experience in much of the ocean today. So even if the water level did reach the heights of our modern mountains as some evolutionists insist existed at that time, the water would have been warmer than the air at that height, thus warming the atmosphere. So even if the water cooled they only had to be in the cold a year at most in the evolutionists' freeze-theory, because the Ark landed about a year after the flood began.

Oxygen

Another situation that evolution supporters attempt to assert is that because the elevation would be so high when the water covered the maintains, Noah and crew would not be able to breath due to the lack of oxygen that we experience at high mountain altitudes today. But this argument is severely lacking in logic because if the water did actually cover the modern-day mountain heights, the lower elevation atmosphere would be displaced by water, thus forcing it up along with the water's surface. And we can prove this by filling an aquarium with air and then pouring heavy gases into it, such as sulfur hexafluoride. You can float a lightweight aluminum foil dish in such a gas and if you were to pour water in the aquarium, then the sulfur hexafluoride would sit atop the increased water level. This is the same thing that would happen to the entire atmosphere. Noah and crew would have had no shortage of typical day-to-day air. Then to add a thought to the last section here, the air temperatures would also have followed the average sea level. And that is all besides the fact that our current-day mountains did not exist then.

Some people will claim that without trees there would be no oxygen, but this too is illogical. Since most of our current-day oxygen comes from the oceans, there would have been plenty for all of the passengers on the Ark. Besides that, there would have been no other creatures alive to use up the existing oxygen. This is a big Earth and if they only have Earth's available oxygen at the time of the flood, and no more was ever produced, they all could still have lived on Earth for a very long time.

As for the available oxygen on the Ark, we can safely assume that it might have been quite smelly during the forty days of rain but, they would have been able to open the window and possibly the door and have had plenty of fresh air. We don't really know how tightly the Ark was sealed, so there might have been plenty of ventilation between logs or even vents added for air circulation. This of course, could be considered to contradict the point made about a tightly sealed Ark that could withstand extreme cold and keep the crew warm enough to survive. But as mentioned earlier in this section, the atmosphere would have risen with the water higher than the mountainous altitudes which would no longer be considered high, and that along with the increased water temperature would unquestionably allow for warm summer-like temperatures even if the mountains were miles high if the water covered them, which obviously would not have been miles high and will be addressed in the chapters ahead. Also, all of the body heat produced from the animals would have demanded ventilation to keep the passengers at a reasonable and safe temperature with plenty of fresh air throughout the entire voyage.

The Ark and its occupants would have had plenty of quality air, warmth, and food to easily survive for as long as they needed to. The Ark would easily have robustly supported them firmly on the surface of the single vast worldwide ocean for many, many days, even years if needed.

Chapter 16

Land Ho!

As the crew of the Ark waited to see dry land appear, Noah was releasing birds to see whether they would return. If the bird returned, then it is obvious that there was likely no dry land yet to be seen. However, if a bird did not return, then that bird might have found a place to perch. Noah repeated this several times until finally one day a dove he sent out returned with an olive bough or a leaf signifying that there was vegetation somewhere out there.

But would it be possible for an olive tree to grow quickly enough to a point where a dove could actually take a bough or leaf from the tree and return it to Noah? After all, the flood covered the highest mountains by fifteen cubits. There are a few key points to be considered here in order for that to be a viable possibility.

Land Sinks and Rises Quickly

To try to understand how an olive tree could have grown so quickly, we have to ask how long of a period of time the water had covered the highest mountains by fifteen cubits. You will have to judge the following proposals for yourself to see they are fair to consider.

Since it does not specifically indicate how long the highest mountain was covered by fifteen cubits of water, we can, without cheating, say that it only needed to be minutes before the water began to recede. Now, of course this is a somewhat ridiculous thought because the water is not going to evaporate that quickly under any circumstance.

But consider these following possibilities as valid points: The first is that if a wave or an ocean swell increased the water height any amount, then the level of a very calm ocean could have been several feet lower than fifteen cubits over the highest mountain top, but the ocean swell could have briefly been fifteen cubits over it. Now, if you think that is a bit sketchy, then consider the tide caused by the moon. Since the tide is a considerably slower ocean swell covering a massive area of ocean, it is a fair and viable theory to use in order to help to reduce the amount of water needed to flood the entire globe with fifteen cubits of water over the highest mountain.

In addition to tidal ocean swells, and then the waves in addition to that, there is also the possibility that the highest "mountain" had fully grown olive trees on it during the flood—trees that never needed to be under water even during tidal and other ocean swells. This is not trying to state that this is what happened, but rather how things could have transpired to allow an olive bough or leaf to be available that quickly. A well rooted olive tree possibly could have had its top completely out of water and only temporarily ever covered by tidal ocean swells.

Regarding olive trees, we also have to realize that while the highest mountains were covered by fifteen cubits of water, it does not necessarily mean that an olive tree was *ever* covered by fifteen cubits of water. A tree is not a mountain; rather it is something that grows on the mountain. Modern-day olive trees are about twenty-five to thirty feet tall. Using our cubit length for constructing the Ark models, this would make fifteen cubits to be about twenty-five feet. So theoretically, the olive tree could have been sticking out of the water about five feet the whole time using today's olive tree standards. And since we don't know the growing conditions of Noah's time, the olive tree could have been considerably taller. There are so many possibilities for an olive tree to be producing green leaves within the year of the flood that these points have to be acknowledged as viable possibilities. Further, the olive "bough" might be a more poetic translation; it is possible that it was seedling growth, which technically could be considered a "bough". Further, some Bible translations say "olive leaf" rather than "bough".

A modern olive tree can produce fruit in about three years. Now unless the dove's olive branch had olives on it, the tree only needed to be a seedling in order for the dove to pluck anything from it. Based upon the geological activity that would have had to occur due to pressures and friction from continental shifting, the climate temperature would have been like the climate of the tropics. And if volcanoes had at all erupted and spewed ash into the air or water, the soil could easily have been perfect for robust growth in a year's time. All of the suggestions just laid out are realistically viable, even if some are a bit of a reach.

But let's also consider geological activity. If the tectonic plates collapsed from the rock matrix crumbling beneath them, then there would have been on-going activity for quite a long time afterwards with months, years, decades, or even centuries of aftershocks. Since it is blatantly obvious that the mountains we witness today were created due to geological activity, it is undeniable that after the flood reached its heights, some level of tectonic activity would have occurred. When we see that our current mountains have risen up above the plains level surrounding them, it is clear that tectonic activity in Noah's day could have rapidly pushed land up at any point after the fifteen cubits was achieved. As has been witnessed in our modern era, land can rise and fall quickly when there is heavy earthquake activity. The land could have risen above sea level immediately after the water stopped flowing. If an area of land did rise quickly, it would have allowed new vegetation to sprout up within weeks of the ground being exposed.

Within a couple of months of an olive sprout appearing, the dove could have easily had her pick of vegetation to return to Noah with. Is this a viable possibility? Yes, it is almost a certainty. However, there is no way that enough water could **_ever_** have evaporated into the sky to remove even a few feet of the fifteen cubits of water above the highest mountains. So, what might have happened in order for any land to appear whatsoever?

The land had to have shifted and pushed up through the water from seismic activity, so the question is not **if** the land shifted, but rather **when** it shifted. Often when these sorts of debate points are offered, evolution supporters will try to find any way they can to disallow each point. However, each point made here has a far greater likelihood of having occurred than even the first evolutionary starting point suggested by any evolutionist, ever. Even Darwin, in one of his books, spoke of land elevation of an immediate area rapidly changing during geological activity.

We also have another point to consider, which is that since the continents would have had to have collapsed in a somewhat short period in order for all land to be flooded, a fair amount of land could have been out of the water for most of the 40 days. If that were the case then when it finally collapsed shortly before the 40 days ended, it could have absorbed some amount of water once it became submerged. This is similar to suddenly forcing a sponge under water. Initially the water level will rise an amount equal to the spatial volume of the sponge, but then the sponge will quickly absorb water until it reaches full saturation and while this is occurring the water will be receding or going down again.

The only thing stopping any of the points just made from being fully accepted and embraced, especially the land being rapidly raised up, is evolutionists' insistence that all continental tectonic movement took millions of years to occur. But even in this, the evolution long-age view fails to recognize what even Darwin himself witnessed and acknowledged: With geological activity, land moves, and when it does, it sometimes will quickly lift up or drop down.

Subduction Myths

"Subduction" is horizontal movement of some of Earth's crust as it theoretically is forced down into Earth's mantle below a neighboring tectonic plate or the sea floor. Subduction has been moving at a rate of about five centimeters per year plus or minus a few centimeters, or about 2 inches per year. For a mile of movement, it would take about thirty-one-thousand years plus or minus about fourteen-thousand, according to some estimates.

The biggest problem that we face with science today is that we incorrectly believe what we see today is what has always been. Since the subduction movement rate for a mile takes about thirty-one-thousand years, we take that figure and multiply by the visible two-thousand miles of movement we see that occurred between the mid-Atlantic ridge to the coasts of the contents and arrive at a figure exceeding sixty-million years for the continents to get to their current locations.

While "subduction" is obviously occurring to some extent, we have to realize that the movement rates we see today are most likely not as they have always been. When we choose to assume long ages that are over sixty-million years, it changes our assessment of the rest of the evidence.

Along with this myth of long-age subduction comes big bang theology. This all gets jumbled up into a big get-nowhere debate between big bang and six-twenty-four-hour-day Creation. Often, people are at polar ends of this Earth-age topic. Could God have created the Earth in just a few days? Not likely. Did the Universe and the Earth bang into existence? Also, not likely. Then what did happen?

The answer is that we simply don't know the actual age of the Earth, but it is very old. God is not going to cheat the grand Creation and the physics derived from it. When we see light coming from distant galaxies, we know that those galaxies are very old. While we can only estimate those distances with sketchy redshift theories, we can be certain that they are very far away. To be sure of this, we have a very accurate estimate of the speed of light and we can prove those estimates to be true when communicating with space probes. We can also gauge the distance of nearby stars using triangulation and parallax.

We can even observe far away galaxies. When we look into the heavens with our telescopes, we see galaxies, and then when we build an even more powerful telescope to see yet deeper into the heavens, we then see yet more galaxies, and when we build an even more powerful telescope, we see even deeper into the heavens and see still more galaxies. This is unlikely to ever end no matter how deep we look into the heavens. What this proves is that those galaxies are very far away and the light would have had to have been emitted ages ago in order for us to see it today. Basic simple logic forces us to know that the Universe is **very** old.

For the reasons just mentioned, we can also safely assume that our Earth is very old, and because of those accurate assumptions we also then continue to make far reaching assumptions and look at the subduction occurring in various areas on Earth and assume that the movement occurred over long periods. This is where science goes wrong and begins to invent illogical and unprovable theories. And it all comes down to the layers of sediment that we often refer to as the "geological record".

Anyone who has taken adequate time to study the various landforms around the world will know how diverse the topography and geological evidence is. We can see it with our own eyes and hear logical and accurate theories, but then still commit to illogical theories that could *not* possibly have occurred based upon the wide range of evidence found all around the world.

Chapter 17

Evidence is Everywhere

The flood topic is the single greatest fear of evolutionists. As mentioned earlier in this book, when the layers are proven to have been deposited within the last several thousand years, it completely destroys the entire evolution house-of-cards. As we look all around the Earth, we see a certain consistency regarding the layers of sediment and the creatures caught in those layers. But when it comes to man, we have yet to find evidence of man in any lower layers in which we have thus far searched, though that point is debatable when considering some of the archeological finds in the various mining industries.

The evidence is trying to tell us something, but we won't listen. The creature forms have been transferred from creature to fossil, making a wonderful record of events that the rocks are trying to share with us by speaking out in the form of logical evidence–If only the rocks could actually talk!

People Migrate

We have clear patterns of the migration of man, but that migration trail does not explain the anomalies. The explorers of the Middle-Ages and of the Renaissance era traveled the world and began to inhabit distant lands. But the native populations do not follow the explorers' and settlers' migration patterns. Man is now and always has been quick to invent and adopt useful technologies, such as boats. Every culture that has a large body of water in close proximity has boat-making well mastered to suit their own purposes. When you consider distant islands that are so far out in the middle of the pacific that they are not visible from any of the main lands, and also consider that people are found on those islands, it is really quite astounding. Did these people get lost at sea and find land and then settle there because they were trapped? We may never know for sure, but what we do know is that man is resourceful and will always find ways to survive. It might be through migration when climate conditions worsen, or through wise management of resources, but man will survive.

When the flood came, everyone, except Noah and family, was drowned, so any current native tribes from around the world had to have come from Noah and family, according to the Bible. If the continents quickly drifted immediately after the flood, then Noah and family would not have had enough chance to increase and multiply and be on separate moving continents, thus if the land moved rapidly then there would have been no one trapped on it to begin a culture of their own. This is why the land being divided when Phaleg was born is the most logical answer to the distant migration situation.

If Phaleg was born about one-hundred years after the flood, then Noah's son's and their families had plenty of time to have several children and then those children have children etc. Noah to Shem to Salah to Eber to Phaleg. Phaleg was three generations after Shem, and Shem had two brothers who were also having children. By the time Phaleg was born there could easily have been many hundreds of people on Earth, even thousands. That's enough for people to be able to migrate and get trapped on a drifting continent. There are obviously many other migratory possibilities, but this one makes a lot of sense when considering the unique nature of some of the early inhabitants of distant lands together with the Bible's Phaleg comment.

Cultural Clocks

As recent centuries' explorers ventured out, they found ancient ruins, but they had no way to interpret the communication glyphs that they often found. After a fair amount of studying, archeologists and researchers began to untangle some of these ancient languages. In their attempts to interpret a language, researchers must make some assumptions about the meaning of the various glyphs found. In one such analysis, certain glyphs were believed to indicate a "base-sixty" numbering system, rather than our base-ten numbering system. Through this base-sixty numbering system and based upon an ancient calendar found, researchers believe that the ancient civilization understood the Earth to be on a twenty-six-thousand-year cycle implying that the ancient civilization was very old. If this is true, then it is potentially a serious problem for the Biblical flood story. It may be true that Earth is on a twenty-six-thousand-year cycle; however, if the researchers are not correctly interpreting the language, and were off by a factor of ten, then the twenty-six-thousand-year cycle would become a twenty-six-hundred-year cycle.

There are many such findings around the world with creative interpretations of time. If you mess with the timeline to achieve your own agenda and insist that a certain number of years transpired within a particular culture, you can then creatively invent any timeline you choose. However, if we toss out all of the long-age timelines that are highly speculative, and instead focus on the events depicted in drawings done by those cultures, then things typically align with the Biblical flood account timing very comfortably.

The Advent of Languages

When considering world languages, we have to realize that it is likely that none of the ruins we discover in our modern era are from before the flood. It might be possible, but it is unlikely that any ruins found today are from before the flood. The pre-flood world probably had a good amount of sophistication to it and likely had well-developed societies, but all of that was washed away and covered in silt when the flood came—and all of the layers together are over a mile of thick in some cases. Noah and family would have had to start all over again. Not that they would have had to reinvent language, but rather all language would have come through them. In Genesis Eleven it says "And they said: Come, let us make a city and a tower, the top whereof may reach to heaven: and let us make our name famous before we be scattered abroad into all lands. And the Lord came down to see the city and the tower, which the children of Adam were building. And he said: Behold, it is one people, and all have one tongue: and they have begun to do this, neither will they leave off from their designs, till they accomplish them in deed. Come ye, therefore, let us go down, and there confound their tongue, that they may not understand one another's speech. And so the Lord scattered them from that place into all lands, and they ceased to build the city." This is one explanation for the various languages, but with language being the malleable function that it is, it has since morphed into many other sub-languages.

We see evidence of this where there are certain languages that are a base-language from which other sub-languages have obviously arisen. When we teach our children well, then we can keep a pure language alive as long as we are persistent in our diligence of preservation of that language. However, once we allow our children to deviate from the language using slang-type terms, then a language can devolve quickly as was witnessed in America shortly after the turn of the twentieth century. The negligence and abuse of the English language ushered in many new terms as well as deviation of definition from existing terms. The new sub-languages are easily understood by those adhering to the pure language. However, if such language perversion of the pure language occurs inside of single generation and then those who spoke the pure language die, then the new sub-languages will become the standard and will be taught to subsequent generations.

The lack of language integrity that was infused by the youth of the early twenty-first century will likely be similarly perverted in a new sub-language as their children come of age and add new terms and redefine old terms. If you do this for a few generations then you can end with a language that would be unrecognizable to those who spoke the original pure language.

Many people believe that we evolved and at one point were ugha-ugha-cave-people, but we have no evidence that this is so. Everything we find clearly demonstrates that man appears suddenly, as well as the developed languages along with man. The Genesis Chapter Eleven story about Babel is where we get the term "stop babbling" due to the language confusion that God caused to happen to the people because of their arrogance in trying to build a tower to fight against Heaven (but that's another topic). If you can accept that story as an account of what actually occurred, then you can clearly see how things unfolded regarding migration and languages and sub-languages. And since Phaleg's birth was about one-hundred years after the flood and the tower of Babel was about one-hundred-thirty years after the flood, the timing works well regarding migration and the land dividing by slowly drifting regarding the languages of the various distant cultures.

Some of the stories in the Bible lack the detail that can be found in other complimentary ancient writings. If translators didn't have the additional information found in these complimentary writings, or if they reject it because "it's not part of the Bible", then they will interpret and therefore translate the Bible as they choose to understand it.

Much the way Genesis One's Creation account has been very badly translated in many late post-Reformation Bibles, so too are many other parts of those Bibles poorly translated. It is those poorly translated phrases that make the Bible appear to be fairytales in the eyes of many people. If God exists, then realize that these misinterpretation issues are not pleasing to God.

Chapter 18

We Know what We Want to Know

Our human biases have always been a problem in all facets of life. We want what we want, and it's often at the expense of others, and sadly, it is often at the expense of the actual truth. The same is true with interpretation. Because we view everything from our own little mental fortresses, we often reject what other people say without ever actually evaluating what they are proposing. If someone pushes back against some theory that you have, then realize their resistance is not a test of your fortitude to stand by your theory, their resistance is a test of the fortitude of your theory itself. If a theory cannot withstand scrutiny, or if you will not allow the theory to be scrutinized, then it is not much of a theory at all. Someone might say, "Well, that theory is preposterous!", which might appear to be so, but the theory might actually be true. The main reason a true thing can appear preposterous to another person is because they have accepted false or wrong information as "fact", thus skewing their perception of reality.

We know what we want to know because we are comfortable with our own beliefs. There is nowhere that this part of our human nature is more dominant than in religion, including God-based religions **and** science-based religions such as evolution. There are many very good people out there debating either side of the evolution-versus-Creation conundrum, but most uninvolved people have a general opinion which they hold silent and are waiting for something that they can grab onto that cannot be legitimately dismantled by either side.

We do what we do, and we desire what we desire, and as we gather our harvest of data, we tend to have our way with it to force that data into submission to meet our own desire of what the data means. But that won't make it correct or accurate or possible when it is not actually correct or possible.

Prolific Spread of Humanity

In the centuries before the flood occurred, man lived very long according to the Bible. Living hundreds of years would allow a longer reproductive period; however, if women had similar cycle rates as is experienced today, then their lifespans would not have had much effect regarding increased reproduction rates. If girls are born with limited eggs and they cycle monthly, the women back then would have been past child bearing years at about the same age as we see with women today. However, life longevity would have encouraged their attempts to have more children at older ages if the people did actually live for *hundreds* of years. Normal, modern-era individual reproductive rates alone are more than enough to quickly populate the world, especially with grandparents and great grandparents living so long and being able to assist the younger generations in safely bringing up children. Even today, many people would have larger families if they had more assistance with managing the children.

All archeological evidence points to very fast proliferation of man. Yet, if we are going to take the Bible seriously, we still have to consider that there are two worlds, the old world before the flood, and the newly purified world that arose after the flood. The question we must grapple with is, which world of man is our archeology revealing to us? Is it the old pre-flood world? Or is it the new post-flood world?

Stone Age

If you look back at the strong evolutionary indoctrination that occurred during twentieth century, you will notice that it was assumed that people lived in caves during the so-called "Stone Age" and that man was rather primitive at that time. If humans evolved, then this would be a very logical conclusion, because if man evolved from some primitive animal form, then we would likely have lived like animals. However, since we are allegedly descended from primates, it is more likely that we would have built actual nests made of twigs and leaves like some other primates do.

No matter how man came to be, we still would have been faced with the effects of starting with nothing. If man is truly created in The Image of God, then our "instinct" would be similar to God's instinct. If you have ever observed a bird building a nest you will know that it's a pretty amazing thing to watch. This tiny little bird, only about a half year old, hatched and grew and now somehow knows how to build a nest, lay eggs, sit on those eggs to hatch them, and care for the chicks afterwards. How does that little gal know that she needs to sit on those eggs to warm them? How does she know that she must bring food to the chicks after they hatch? And how does she know when the chicks are old enough to be forced to learn to fly?

Just as animals have instinct, so does man, but the instinct of man is elevated considerably above that of the animals. Is our "instinct" learned or is it natural? Some instinct is natural for sure, like if a mother holds her newborn infant close enough to her breast, then through a simple gentle touch of the child's cheek the child will quickly find the mother's nipple from which to acquire nourishment. But how did women ever figure out that they needed to do this? Was it from watching the animals? Was it an evolutionary trait passed down from when we were allegedly animals? If we accept the vast speculation offered by evolution theology, then it is easy to buy into that belief, but it truly doesn't reconcile with reality. All of the evidence we see shows that man almost immediately lived in "manmade" structures of some sort. Of course, there is evidence suggesting that some people lived in caves, but we have no way of knowing for how long they did so, or how long ago that might have occurred.

Imagine for a moment that you live on a distant secluded tropical island and have a generator and all of today's modern luxuries of life. You and your family are enjoying a casual life of sunshine and abundance everyday living in total comfort! You even have animals and pets and abundant fruit trees. As you explore your private island, you discover a cave and all sorts of interesting features on the island. Now a powerful hurricane is headed your way. Where are you going to go? Maybe you will go into the basement of your home if it has a basement, but the hurricane rips your house clean off its foundation and you are now looking at a violent sky, so you huddle in the corner for days until the storm passes. Now where are you going to live? If you had not made your way to the cave immediately after your house was ripped from its foundation, then you certainly will afterwards until you rebuild your home.

It is unlikely that people inhabited caves any longer than they had to. Caves were probably only used when traveling, or in emergency situations, or when starting anew. Caves are strong and are typically more constant regarding climate. On cold days they are warmer than outside, and on hot days they are cooler than outside. Caves offer quick, free, effortless accommodations. Our caveman imaginings are just that—imaginings. Explorers find caves with ancient drawings and assume that those drawing were the best people could do during the so-called "stone age", but while that is possible, it is also a possibility that it was a sort of graffiti of the times—It might be that someone was simply expressing their artistic flair. They also could have been drawing things that they wanted to show people far off in the future. We simply do not know for certain the reason cave drawings were made. What will people make of the artistic works from our modern era four-thousand years from now?

The "Stone Age" is named so because of the stone tools found surrounding many areas where artifacts were found. This would be the same sort of stone tools you would find yourself making for survival on your private tropical island after all of your possessions were lost in the hurricane. And over time, you would perfect those tools to make your family's lives easier.

Neolithic Age

The so-called "Neolithic Age" is that time during the latter part of the "Stone Age" when the people began to perfect their tools by grinding or chipping away the stone to shape it for their specific utilitarian purposes. If you study the consensus on the timeline for the various "ages", you will find that the Stone Age allegedly lasted about 2.5 million years, coincidentally only ending about five thousand years ago when man figured out how to work with metals.

Bronze Age

The "Bronze Age" is perhaps the ultimate key to either prove or disprove the theory that man evolved from source primates. The term "primate" comes from the term *prime*, basically meaning *first* and nothing more than first. The term "primate" is used for other purposes than evolution's hijacking of the word. The Bronze Age is alleged to be the time when man figured out how to melt metals and shape those metals for needed purposes. One primary indicator that man is relatively recent is the sudden arrival of the Bronze Age. Interestingly, the evolution theorists have conveniently made the Stone Age last for about 2.5 million years right up until the time of the flood. This is done because the evidence is so overwhelming that metal tools began within a couple of hundred years of the flood. Creationists imagine that Noah and family would have figured everything out again all by themselves, but this is unlikely. While they might have had to do everything from scratch, it does not mean that they had to figure it all out **again**. The Earth was cleansed of iniquity, not knowledge.

What would have occurred after the flood is that they would have had to find sources for the copper and zinc or tin needed to make rudimentary brass and bronze. There would not have been any known nearby sources of any of those metals immediately after they stepped off of the Ark. Their first concern would obviously have been reestablishing the basics of life. They could easily have lived in the Ark for the first several years and then as the waters receded more and more they would have worked to establish new dwellings. Noah most certainly had some tools along on the Ark. But those few tools would likely not account for any artifacts that we have unearthed up to this day.

It doesn't take man long to find things. After a couple generations there would have been hundreds of people living their lives trying to improve life just as we all do today. The only difference with today and back then is that things would have been a bit more basic back in their time than they are in our modern times. The "caveman" bronze-era theory has too many faults in it to be taken seriously. No artifact found fits properly with the bronze era caveman theory, yet everything except the alleged and wildly inaccurate timeline that the evolution-theorists attempt to force into our origins, fits perfectly with the Bible's flood account when you understand the flood account properly.

Noah may have brought some tools along on the Ark, but after you have a couple of hundred people using those same tools, the tools are going to get worn out, broken, or altogether lost.

And while one person is using those same tools other people would also have wanted to move forward with their lives by using the tools. It is very clear if you think this through logically, that the first couple of generations would have had limited resources immediately after the flood, just as you would on the remote private tropical island after a hurricane. You would use rocks to get the job done and you would use the cave for shelter until you completed something more accommodating. These are simply logical truths of what would have had to occur during the first couple hundred years after the flood. They would have used stone tools and then eventually have perfected them to a "Neolithic" state ("neolithic" means new stone). After there were enough people around, they eventually found sources of required metals and the "Bronze Age" was born. There is no evidence of a 2.5-million-year-old "Stone Age". There is evidence of a so-called "Stone Age", but the time-scale of the era is not known for certain. Evolution theorists say 2.5 million years and some Creation theorists recklessly assume that any found evidence was from before the flood.

Other Ages

Are the Stone Age, Neolithic Age, and the Bronze Age the only "Ages"? No, there are several other key "Ages" where man made great strides forward in innovation and learning. There is the Egyptian Age, the Greek Age, the Roman Age, the Renaissance Age and the Industrial Revolution Age. But here again we get in to the pesky labeling problem like what occurs with defining species, what makes the distinction between one "Age" and another "Age"?

As archeology clearly shows, there is one "Age" and it is the "Age" of man. All evidence we find is consistent with man arriving suddenly. When Genesis says that Adam and Eve were created, we can believe it right? Not necessarily. Then it was evolution, right? Maybe. There's a lot more to that story than meets the eye (See *The Science Of God Volume 4 - Day Six - Evolution versus Man - In Our Image*).

Chapter 19

Layers of Deceit

The Earth has securely preserved, for us, some of the history of the entire globe. The truth is retained in the layers, with layer upon layer of evidence all stored up for us to discover and compare to our current world experience. Those layers are trying to convey all of that history to us so that we can retain the artifacts and convey the truth of those artifacts to future generations. But alas, all of the stored-up artifacts are often misrepresented to subsequent generations by both sides of the evolution-versus-Creation and flood conundrum.

Layers of Dirt

As mentioned in an earlier chapter, there are groupings of layers of sedimentary rock within larger groups. Some of these layers are nearly impossible to distinguish from one another within their larger group and are only realized upon closely examining edges of a layer-fragment very close up. When the seamline between layers is discovered, you can take a chisel and drive it into the seamline with a hammer, and then typically the layers will separate. Then, often, there will be a treasure-trove of fossilized creatures—usually water insects and small fish. The interesting thing about this is that if evolution is correct about the ages of the layers, then what is the explanation for the intermediate layers that have bonded so tightly together? If we look at items that recent man has deposited on the sea floor like sunken ships etc. Then what do we see in only a matter of a few years? We see a consistent pattern where items under water for even a few years will have sediment and currently living as well as dead sea-plant life covering them. If there was any more than a few months between the sediment layers that we currently find, those layers would easily be separated by that type of matter. In fact, this is what we see today in archeology. The layers that were deposited in rapid sequence are tightly bound, but can be relatively easily chiseled apart. But the greater layers that contain those layers have other types of material between them allowing them to come apart easily when handling them.

The issue of the amount of dirt and foreign matter, or lack thereof, between layers is problematic for both evolutionists and many Creationists, and here's why: The dirt between any two layers within the major strata is not consistent with millions or thousands, or even hundreds of years for that matter. There is not enough dirt between layers for long-age deposition to be a logical possibility, and that is true between any two layers from the bottom to the top of the entire major strata range—for example, the entire height of the Grand Canyon. The obvious, logical, and undeniable evidence against long-age deposition is as abundant as the land that we all walk on today, and only deliberate ignorance will indicate otherwise.

Which Layer is the First Layer?

Regardless of long-age deposition, or deposition due to the flood, we have to consider *all* of the layers. If we are going to claim that the continents "drifted", then we have to acknowledge that they moved across the sea floor. If this is so, then logically, the continents' thickness can be calculated by using the depth of the ocean plus the elevation of the land. In most cases this is about three to four miles without considering the mountains. The question this brings up has to do with the sea floor. Since the flood was worldwide, surely the sea floor would also show signs of similar sediments as found on the continents. And we would expect those layers to very loosely resemble the layers found on the continents. As for the continental thickness averaged with ocean depth and altitude of the plateaus of the continents, we find the thickness to be about three miles or roughly fifteen-thousand feet.

However, we have to take into account that the added continental weight would potentially cause the continent to cause a slight depression in the ocean floor surrounding the continents and beneath them. This could be a depression of measurable distance of hundreds of feet. But we also have to consider that the ocean floor should also have a relatively close thickness of deposited layers depending upon when the continental drift movement occurred, which, depending upon the timing of when the continents moved and when the various layers were deposited it could increase the ocean depth relative to the land by as much as the depth the Grand Canyon, which is about a mile. This means that the plateau areas of the continents might be as much as four miles thick or about twenty-thousand feet thick.

When examining the mountains and other areas that display the occurrence of extreme geological activity, you might notice rare cases of overturned layers almost like folding something like taffy back onto itself. You can think of this in terms of an "S" bend. However, while this sort of thing is somewhat common, most such areas are fairly small and usually include several sediment layers being bent together. The interesting thing about such layers is that they would break rather than bend and stretch to the extents that we see them as today. The only way that rock will bend is if it is soft. Evolution and big bang supporters will claim that since everything took millions of years, theoretically, the rock would have bent so slowly that it would not have fractured. But we know this to be false because rock is very fragile and will not bend in that way. In order for the rock layers to bend as we witness them in the various landforms today, the layers would have had to have been in a putty state in order for the layers to bend *and stretch* around the massive bends even if any layer took millions of years for it to conform to its current state

When we examine the Grand Canyon, we see that at the base of the canyon, the rock is formed from melted rock rather than from sediment. Once the sediment begins in the Grand Canyon, then the layers are sediment from that point upward. If we assume that the flood deposited all of the layers in the Grand Canyon, then we will eventually need to prove that out.

On Top of All That

When considering the typical Creationist's analysis of the layers and Stone Age artifact evidence that is found, it is often attributed to the pre-flood era. And regarding "Ages" of man, there really are only two eras, our Age and the Pre-flood Age. To understand this better, you must first realize that there is very little evidence of human activity found between sedimentary layers.

Almost all, evidence of human activity has been found above the layers of sedimentary rock. This tells us something very important. First, is that this is not consistent with evolution. If early primate-man walked the early Earth, we should have found abundant transitional evidence between some of the upper layers, but we haven't. This is where the timeline card is magically pulled out of the sleeves of the talking-heads of pop-evolution. If you listen very carefully, the evolution timelines are very loosely defined much like species are, and within that, you will find conflict of what came first in the time-line if you look for it throughout *all* of their analysis—the timelines do not all add up properly as they like to imagine them to.

Another thing that the lack of human artifacts between layers tells us is that, generally, the evidence of the pre-flood people would not be found on top of the layers or anywhere near the top layers. Such evidence, if it exists, would be locked in the layers with the animals and man who was fully formed just like modern man. If we were to stumble upon some clearly evolutionary primate-transitional artifacts, then that would confirm evolution, but those transitional artifacts would have to change progressively as you go lower in the layers. And since the animals died in the flood according to flood supporters, we generally cannot attribute the human artifacts that we find near the current land surface to pre-flood people, specifically because the pre-flood people would likely be deeper down mingled *in* the lower layers, not above all of the layers.

Chapter 20

What Do the Layers Tell Us?

Why are the layers of sediment getting so much attention in this book? It is because the layers are our evidence of what happened. Whatever happened doesn't matter regarding the evidence we find trapped in the stony layers, because none of that will change. But that evidence is there screaming at us to wake up and see the truth. The layers tell us so much, and the battle between evolution and the Biblical worldwide flood of Noah's time will ultimately be settled by those layers.

Progression of Layers

Long-age evolutionists insist that the layers took hundreds of millions of years to deposit. And many Creationists believe the flood deposited most layers within the flood year, roughly forty-five hundred years ago. Is either of these views correct? Probably not.

The motion dynamics are going to be incredibly complex to untangle. If we find a major contradiction in our movement and deposition theories, then we must immediately stop and reevaluate our theory. The typical Creationist flood theory matches the evidence and logic far more closely than does its evolutionary arch-nemesis long-age-deposition-theory. Evolution manipulates the time-line with long ages in order to force their evolution theory into the layers, but **all** physical evidence and logic defy that long-age theory, especially if we accept the idea of long-ages for all the reasons thus far mentioned in this book. Long-age sediment deposition is a self-annihilating theory. The longer the speculated time between layers, then the more logically and visually **not** likely it is that there were long periods between layers, because the amount of variation caused by erosion would be substantial, which it quite evidently was not.

Extreme Chaos

If we did not see clear evidence of continental movement on the ocean floor, it would change the logical assessment of where the "fountains of the great deep were broken up". But since the continents somehow moved, it is logical and likely that there was a supporting rocky matrix beneath the continents that gave way as it crumbled and was largely pulverized into fine silt when it was forced out along with the water from beneath the continents. If not for the evidence of movement on the ocean floor, we could conclude any irrational theory.

What is most important to understand about the flood is that the new topography of the Earth would have been unrecognizable to Noah and family. So, if the flood is true, then nothing would have been familiar to them when the door of the Ark was finally opened, everything from the past would have been buried to a great extent.

If you consider the devastation that a major modern earthquake can cause, you might begin to understand how violent some of the activity around the world would have been during and after the flood, and you will also become aware of how long the violent activity might have gone on. If an earthquake happens in our modern era, we measure it with the "Richter Scale" and when an earthquake is at the extreme end of that scale, then severe devastation will occur, toppling everything from small homes to tall buildings and causing fissures ranging from cracks in the pavement to vast deep crevasses in the Earth. Then shortly after the quake has ended sometimes a powerful tsunami will occur carrying with it lots of dirt and debris as it rushes across any land within its reach. Our modern experience of a severe earthquake is meaningless in comparison to what you would have felt when the continents began to move during and after the flood.

Let's assume for a moment that the flood occurred as implied throughout this book. And let's assume that there was a fragile supporting rock matrix that began to crumble sending pulverized rock mixed with water into both the air and the ocean. Now consider the power of the earthquakes that would have occurred during and after the forty days of flooding. Not only would there have been powerful earthquakes unlike anything we have ever experienced in the past several thousand years, but there would have been massive powerful tsunamis rolling across the ocean, which, according to the ocean floor maps, would have been one big ocean at that time. Tsunamis can range in speed from about forty miles per hour up to several hundred miles per hour. Assuming the size of the Earth has not changed and the land masses were once joined at the Atlantic ridge, the Pacific Ocean would have been something like sixteen thousand miles across. Additionally, take note that when the flood peaked at fifteen cubits above the highest mountain, the ocean was twenty-four thousand miles encircling the entire globe

A slow-moving tsunami would take about two weeks to move across a pre-flood Pacific Ocean, and close to four weeks to encircle the globe. Then at the high-end tsunami speed of roughly five-hundred miles per hour it would take about thirty-two hours for the tsunami to make it across a pre-flood Pacific Ocean, and forty-eight hours to make it around a global ocean.

What we would expect from this activity would be that there would be tsunami waves filled heavy with pulverized rock in the form of relatively fine silt, and those waves would have oscillated across the ocean at varying speeds depending upon the magnitude of the many tsunamis caused by earthquakes that logically would have occurred coming from the continents falling as the supporting rock matrix structure collapsed beneath them. There is no good way for us to know what part collapsed first or the specific direction of the tsunamis or the magnitude or speed of those tsunamis. What we do know is that something carried a lot of sediment around the Earth causing similar layer patterns to appear all around the world. The patterns would obviously not be identical because there would have been regional earthquakes and tsunamis and some tsunamis would have been dampened from colliding with a competing tsunami. This vast array of hydrodynamics would have gone on for years, nearly constant in the early days but calming as things began to settle into place.

Deposition Rate

While there would likely have been massive Earth-shaking activity during all of the forty days of flooding, it would really depend upon your altitude and how far inland you were in regard to the Ark's survival. But when we consider the amount of violent activity that would had to have occurred, it is consistent with what we see in the deposited layers in places like the Grand Canyon. The deposition rate would be extremely heavy and would taper off as a tsunami moved across the ocean and across the land. The force of a wave would change as the silt suspended in the wave began to gradually drop out of suspension as the tsunami waves moved across the face of the Earth.

If you have ever had the opportunity to watch video of torrents of water ripping through the land, then you should be well aware of how thick with dirt the rushing water becomes. You might also have noticed that massive boulders five to ten feet in diameter are pushed around in such torrents as if they are nothing. The sediment being carried in moving water will drop out quickly as the water moves along its way, and if the water could come to an instant standstill nearly all of the heavy larger sediment would drop out within only hours, leaving only a mud colored water rather than a mud thickened water. The layers that were deposited during and after the flood happened fast, and many were the result of tsunamis, but the story doesn't stop there.

There is ample evidence of ancient dead or stagnant volcanoes all around the world, along with the obvious active volcanoes we see in our modern era. Many of these volcanoes would have been very active during the flood and even long afterwards as they were spewing ash, molten rock, and critical minerals into the air. The ejected volcanic debris would have had a few ways of being distributed. First with the massive geological activity, the volcanoes would have been very active and likely very powerful. There would have been volcanic debris drifting in the air for hundreds of miles from the ejecting volcanoes. Any volcanoes in the forty days of rain would likely have not had their material travel nearly as far when being rained on. Then we also have to consider that the volcanic debris would have been getting mixed in with the silt carried along in the tsunamis, thus creating mixtures with varying concentrations of the different materials. This would produce exactly what we see today.

Scientifically there is no question whatsoever that if there was a supporting rocky matrix, that the sort of events just described would occur and there would be no perfectly dependable way for us to calculate the order of events. We might come close, but natural events have a very random nature to them, thus producing far too many variables to be able to calculate them all with pinpoint precision.

The deposition rate of the material temporarily suspended in the tsunami waves would be rapid. The pulverized material and the volcanic material would stay suspended longer while the waves were in motion, but the suspended material would be dropping out of suspension throughout the entire trip across the globe. Additionally, each wave would oscillate in the ocean coming back across. All of this is consistent with all of the evidence that we see in the layers of rock. And most importantly, none of the layers are laid down consistent with long ages.

Chapter 21

Before and After

When Noah and family left the Ark, they would have initially seen water everywhere. When they first ran aground they might not have even realized that the Ark bottom touched ground depending upon how rough the waters were. If using the design of the Ark explained in earlier chapters, the Ark would have been very stable and wouldn't move much from small waves, so they could easily have settled onto ground as the waters receded without them being aware of that. In other words, the Ark likely did not "run aground", but rather may have gently settled to the ground as the water receded. And since the Ark would have a sizable portion of its base raft below the water's surface, they may have been sitting on ground for months until the water receded enough to expose the ground on which they first came to rest.

The old world that they left behind was now forever lost. You have to try to get into the heads of all of them. They would have been stunned at the difference after the flood upon opening the Ark door, versus what they saw just before entering the Ark. If

God didn't give them details of the result of a massive flood, they might have fully expected to see familiar sights from the pre-flood world. Some of them probably had thoughts of being able to go back to the old world and their homes, but that was not to be. Their shock at realizing the magnitude of the flood must have been intense!

Four Corners of the Earth

In their interpretation of the Bible, some translators have used the term "corners" when referring to the four cardinal directions of *east, south, west,* and *north.* "Corners" is one of those poorly translated words included in some translations. Some translations use "quarters", but even that is a poor translation. The original root-word intended for it to mean the extents of the cardinal directions, meaning to be all inclusive of the Earth. When we read these poor translations, we then mistakenly begin picturing things like a flat Earth. You will often hear evolution supporters mock the people, especially the church of Columbus' time, because some of the people from that time allegedly imagined that Columbus' ship world sail right off the edge of a flat Earth. This sort of inept interpretation was occurring even before the Reformation era, so in those days it wasn't due to some of the late post-Reformation Bibles that have been so horrendously translated; rather it was due to the personal interpretation done by the misguided clergy and elders who read the text, much like evolution and big bang theorists today read that same text.

The scientists of those days were typically church leaders and/or elders, and apparently, they didn't use logic in their interpretation of parts of the Bible. It is the incorrect interpretation of the Bible that caused people to believe that the Earth was flat and that it was the center of the Universe. And unfortunately today, there are still people who choose that foolish flat-Earth belief. Although, when searching that topic, it is obvious that there are nefarious actors deliberately trying to deceive the vulnerable into believing that the Earth is flat. In

Columbus's day they imagined that Columbus would sail of the edge of the Earth and that Galileo was misunderstanding what he was observing regarding Earth traveling around the Sun, rather than the Sun traveling around the Earth as his counterparts imagined it. Many scientists today are like the pharisees of Jesus' day or the scientific clergy of Columbus' and Galileo's times. The elders believed what they were taught and what they subsequently interpreted from the scriptures, and no one was allowed to tell them otherwise even when some of them knew that they themselves were wrong.

The tsunami waves that carried the vast amounts of rapidly deposited sediment carried that sediment all around the world from all directions all during and after the forty days of the flood. Things probably calmed considerably after the forty days ended, but major disturbances likely continued to some extent for quite some time.

A Different World

Even if anchors tethered the Ark to the hill that they built the Ark on, they would still probably have not recognized their new home. When they first touched solid ground, the bottom of the Ark would likely have been well over ten feet below the surface of the water, even up to 22 feet. This means that when the very bottom of the Ark finally came to rest on the ground possibly up to 22 feet of water had to vanish **before** they could even think about setting foot on dry ground. Now if you pay close attention to the text concerning the landing period, you will notice something very amazing "And after that forty days were passed, Noe, opening the window of the ark which he had made, sent forth a raven: Which went forth and did not return, till the waters were dried up upon the earth. He sent forth also a dove after him, to see if the waters had now ceased upon the face of the earth. But she, not finding where her foot might rest, returned to him into the ark: for the waters were upon the whole earth: and he put forth his hand, and caught her, and brought her into the ark. And having waited yet seven other days, he again sent forth the dove out of the ark. And she came to him in the evening, carrying a bough of an olive tree, with green leaves, in her mouth. Noe therefore understood that the waters were ceased upon the earth. And he stayed yet other seven days: and he sent forth the dove, which returned not any more unto him. Therefore in the six hundredth and first year, the first month, the first day of the month, the waters were lessened upon the earth, and Noe opening the covering of the ark, looked, and saw that the face of the earth was dried. In the second month, the seven and twentieth day of the month, the earth was dried. And God spoke to Noe, saying: Go out of the ark, thou and thy wife, thy sons, and the wives of thy sons..."

How possibly did the water evaporate that fast? If the water was "**fifteen cubits** above the highest mountains", then how could that much water evaporate so quickly? To try to get a grasp of this problem, you have to consider that **the entire globe** was covered with water, allegedly to "fifteen cubits **above** the highest mountains". That is a lot of water around the entire globe to explain away in *any* period of time, especially in only a few months or even weeks. And even if we take from the fifteen cubits by subtracting tides and ocean swells and waves within those swells as suggested earlier, we are still unlikely to see *any* exposed land from evaporation alone. But just for the moment, lets imagine that the "fifteen cubits" did include oceans swells, then even if the water was only a few inches above the highest hill we have to imagine several inches of water *all* around the world being evaporated into the sky. This means that clouds with the equivalent of, oh let's say, just twelve inches worth of rain in them would have covered *the entire sky* all around the globe all at once, which would make things very dark. This very unlikely scenario is also very unrealistic. So where did all of the excess water go?

Modern Mountains

Noah and family were adrift for about a year. If they happened to be on one of the highest places on Earth when the flood rains began, they could have been sitting firmly on the ground until the last few days of the flood.

When fully loaded, the Ark, built as designed in earlier chapters, would not have begun to float until the water came up around it to well over ten feet in depth depending upon payload and cubit-index used to calculate size. This would have allowed them to avoid a great deal of the flood turmoil. We know from the ancient writings that are complimentary to the Bible, that Noah and family lived on a hill or mountain and that it was there that the Ark was built. This would have resulted in everyone else being not on that hill because they left that safe place to go "down" off to the other people and join in the revelry. If you have ever experienced or watched a powerful flash-flood, you will know that the first running water from the rain can come rushing down a hill or high area from zero inches to several feet in a few short seconds.

Since Noah and family were safely tucked away on top of the mountain, the rains would not have affected them, but to those who lived near the base of the mountain, we'll let's just say that if the rain clouds suddenly burst forth with very heavy rains, they would have run into their homes for protection but then never even realized what hit them when the flash flood came rushing down the mountain and wiped away their houses. Since some of the ancient texts imply that Noah was mocked for his perseverance of building an Ark and preparing for the flood, it is unlikely that anyone thought the rains where going to be that bad. Surprise! We likely can safely assume that anyone who knew about the Ark would have been regretting not heeding Noah's warnings as they were gasping for their last breaths.

When "all the fountains of the great deep were broken up, and the flood gates of heaven were opened...", Noah and family where resting comfortably and safely in the Ark sitting on the hill/mountain on which the Ark was built. We cannot assert that the mountain was as high as our modern-day mountains are because if it was then the altitude would have hampered living by placing them in freezing cold temperatures with little oxygen to breath as they constructed the Ark. They might not have been on the absolute highest hill, but if we are going to take the Bible seriously, then we have to believe that whether through ocean swell and waves, or sheer magnitude of water, that the Earth was, for at least a moment, completely immersed in no less than "fifteen cubits" of water. So how did the water evaporate so quickly? And where did our modern era mountains come from?

Here is perhaps the most interesting part about the flood. When we strip away all long-age evolution and long-age geology, including long-age sedimentary layers and long-age continental drift, we are then able to move things along quickly as is logically apparent in the geological evidence, thus allowing us to easily and scientifically explain everything—just as the geological evidence shows.

Obviously, some of the water would have been evaporated into the atmosphere, but that amount of water is insignificant in this situation. Even ten feet of water being evaporated into the atmosphere would not solve our problem, and then we would have to logically deal with an irrational amount of water being held in suspension in the sky. The skies after the flood were almost certain to be normal skies absorbing normal amounts of water, both before the flood began, and at the time Noah released the raven. "And after that forty days were passed, Noe, opening the window of the ark which he had made, sent forth a raven: Which went forth and did not return, till the waters were dried up upon the earth. He sent forth also a dove after him, to see if the waters had now ceased upon the face of the earth." This text is either misinterpreted, or that raven found a place to hang out. It is possible that the raven flew too far out and found no place to rest and so it ended up drowning. But if the raven did find a place to rest, the text sounds as if it was shortly after the rains stopped. However, it also says "and did not return, till the waters were dried up upon the earth."; this implies that the raven did eventually return.

How could any dry land appear in even a few years, surely fifteen cubits of water would far exceed the saturation level of the air, thus, once the atmosphere hits full saturation, it will rain back down and there is no way the entire global atmosphere is going to hold over fifteen cubits of water that was contained on the entire circumference of the globe. And if it could have then everything would still be under water except the very tip top of the highest peak. Using the cubit scale used for building our Ark in an earlier chapter, the water would have been about twenty-five feet over the highest mountains of that time. This is doom for Biblical flood theory!

But for now, let's make some assumptions. First, we will assume that continental movement was fairly rapid, maybe not forty days rapid, but certainly **not** slowly over millions of years. Now, we know for certain that with the heavy seismic activity of earthquakes, we can see the ground raising and lowering many feet in only minutes. In understanding this, it is not all a stretch that things below the water were shifting and the Ark could have been lifted up by the ground rather than the water receding. But that is a matter of scale and perspective like mentioned in an earlier chapter where if the ground is sinking then at the same time the water is rising, so you are not necessarily going to know that the ground is sinking, you might simply assume that the water is rising. And the opposite is true as well, so from Noah's perspective, if a large enough area of land was rising, it will appear that the water is receding because it is receding relative to the land.

It is not a stretch in any way to assume that the Ark was lifted up by the ground regardless of whether or not there was any continental movement. An earthquake could easily push up the ground twenty or more feet in a matter of hours, or even just a few minutes, and quickly enable the Ark to no longer be in the water. This is not implying that it happened that quickly, but rather that it could happen that quickly with no concern or need of any evaporation of any water whatsoever.

Since we are reasonably certain that our modern-day mountains were at some point completely under water due to the fish fossils found in them, we can be confident that they were under water for some time. However, we will add this little caveat; it depends upon how large the fish fossils are, because it has rained fish. Even in recent times the fish raining phenomenon is more common than you might imagine. When water tornadoes or "water spouts" as they are typically referred to, draw water out of the oceans, they often bring with that water many small fish and other small sea creatures, and that is one potential explanation of how fish got up into the mountains. However, we will dispose of that theory because while it is provably possible, it is unlikely for two major reasons. The first is that the fish are fossilized. The fact that the fish are fossilized means that something had to have rapidly covered them at some point and it had to have done so for a long enough time for the sediment to harden into solid rock.

Then we also have the very obvious rocky landforms that make up the mountains. What do these rocky landforms tell us? Because we understand that water will level itself just as we witness every day due to gravity, the silt that is carried in the flood water, then drops out of suspension in a relatively consistent manner. And we know that any rock with readily discernable "layers" that has fossils in those layers, was at one time flat and relatively level from the gravitational pull of the Earth, just like we see the layers so eloquently displayed all throughout the Grand Canyon and in many other places around the globe. Because of the height of the mountains and the angle of the massive and mountainous rocky shards, it is obvious that the mountains once did not exist and were at one time flat similar to the layers in the Grand Canyon.

As you can now see, there is no problem explaining the Ark being on solid ground in relatively short order if we remove the unprovable evolutionary long-age time-constraints that have been indoctrinated into hearts and minds of far too many people for far too long. An earthquake could easily have heaved up a large section of land that would have quickly lifted the Ark out of the water. We have legitimately solved the fallacy of massive amounts of evaporation, and we now potentially have Noah easily on solid ground within a year of the flood without having to cheat reality and actual true science. However, we are still faced with the fact that the air is already saturated to typical levels with normal cloud cover, and Noah might not have had much more land than a handful of acres that pushed up beneath him. So, in this scenario, we are still stuck with an entire globe of water that has not actually gone down in level more than maybe a couple of inches from evaporation.

This is not looking good for those who believe the Biblical flood—nope, not good at all! If the Bible is true, and we refuse to invoke hocus-pocus magical nonsense or imagine that all of the excess water was absorbed into the Earth, then we have a real problem legitimizing the flood. But at least Noah and family are now back on solid ground!

Chapter 22

What the Evidence Means

We don't get to choose what the evidence means; we only get to make a choice as to what we believe the evidence means. When we reject the obvious, we typically commit to the darkness of error, and in doing so, we trap ourselves in thought that is absent of light. The whole point of trying to figure out this evolution-versus-Creation-and-flood dilemma is to illuminate our life with truth. When truth presents itself to us and we ignore it due to the fire of passion that burns within us for our own preferred theories, then we cannot obtain truth and we will always view things through our rose-colored glasses.

Rose-Colored Glasses

Most people will either condemn the Bible, or they will see it through their rose-colored glasses and ignore what it really is. The Bible recounts the foolish, ignorant, and abominable histories of various peoples from the time of Adam through the time of The Christ and briefly thereafter. But it also includes the very important Creation account and Salvation information. And if you read the Bible through those rose-colored glasses and make God out to be some limp-wristed sedate parent, then you will encounter troubles in life. The interesting part of this is that the evolutionists who claim that they do not believe in the Bible, and they together with the Bible believers, both view the Bible through those dangerous rose-colored glasses discussed in the book *Understanding the Bible - the Bible How-To Manual* AND *The Things We Don't See.*

These rose-colored glasses tend to make people see unicorns, lollipops, and rainbows all throughout the Bible. However, only Biblical rainbows can be legitimately interpreted from the Bible. The Bible is truly a gruff book. Christians tend to go with the "loving God" who wouldn't hurt a flea. But so do atheists, yet they see contradictions in what they themselves believe the Bible says, versus what the Bible actually says. Is God a loving God? Yes, but God typically doesn't take people's horrendous behavior lightly. And as is repeatedly presented throughout the Bible, if you don't change your ways you will eventually pay a very heavy price for your own choices. It is hard to determine who is more guilty of sin in this regard; the people who profess to be believers but then proceed to promote lies through their inaccuracies, or is it the atheists who detect the contradictions between their understanding of the Bible, versus what the Bible actually says, but yet fail to correct the record and instead they turn from God and try to turn others away from God along with themselves.

Remove your rose-colored glasses if you happen to use them and cast them far from you so that you do not continue in personal deception, potentially leading yourself and your own family astray. If you want the truth, you will not find it unless you are the type of person who looks at actual facts and examines those facts without your own personal agenda influencing your conclusions. Whatever happened in the time surrounding the flood happened, and the evidence is truly everywhere around all of us and is present all around the world. If you look for the sort of events mentioned throughout this book, you will see them and they will become undeniably apparent to you without having to invent ways that God cheated nature to deceive us into believing the Bible.

Does the Bible Contradict Itself?

When debating the Biblical topics, many opponents will take the relative few perceived discrepancies and reason through them in an illogical manner while claiming that they are being "scientific" about their analysis, but then often go on to completely ignore all of the rest of the text in their "scientific" discernment. If you say that one aspect of the Bible is wrong and then discredit other aspects due to the one discrepancy that you are imagining, then it is on you to prove the supposed discredited parts of the Bible that have "absolutely no merit based upon actual evidence". For instance, if someone cannot prove their case with solid logical arguments and or evidence, then neither can we assume that Columbus's expedition ever made it to the Americas.

There are those who say that the claim of Columbus discovering the Americas is not true. So then does this mean that the Americas do not exist because all accounts of such an important discovery are not in perfect harmony? Or maybe that while people live in the Americas that our arrival is told inaccurately? Which will it be? Is all of the information that has been recorded centuries ago to be disregarded because it has no validity because *some* people do not believe that it is *perfectly* accurate, and thus it never happened? This sort of flawed logic is foolishly deceptive to those who choose to not be discerning about the information being discussed. This is why the books *Understanding the Bible - the Bible How-To Manual* AND *The Things We Don't See* and *Understanding The Church - Upon This Rock I Will Build My Church* are so important for everyone to read. Even if you don't believe the Bible to be true, you are still foolish to speak against any Biblical subject if you don't truly understand it.

The real question that must be looked at with regard to the
Bible is this. Do the places spoken of in the Bible exist or not?
Are there other non-Biblical written accounts of the people
mentioned in the Bible? How much of the Bible can be
substantiated?

The amount of the Bible that can be substantiated is far
greater than most people grasp. But since the Bible has so many
prophecies about future events plus other interesting statements,
then can we assume that it was written after the fact and
invented as some entertaining fairytale? To those with a
inclination to deny the authenticity of the Bible's text, if you are
going to cast away Biblical history as false, then you have to cast
away all Egyptian history, Greek history, Roman history, Hebrew
history, or any other culture that is specifically mentioned in the
Bible, because if the Bible cannot be trusted with all of the
overwhelming physical evidence of places and people and all of
what has been discussed in this book thus far, then neither can
we trust the less documented other Biblically-complimentary
histories just mentioned.

There is much text in the Bible that is somewhat trivial
information that people outside of the matters described in the
Bible would not be privy to. You can think of this as similar to
the secret government details that are discussed in the inner
sanctum of any government, but will at some point years later
likely become public knowledge.

There are no contradictions in either the Creation account or
the flood account. And when we remove the human-imposed
time constraints that evolutionists and big bang enthusiast place
on the timelines, then everything fits together better than
puzzle-like perfection. The time constraints are forced into the
discussion by both sides of the evolution-versus-Creation debate.
The evolutionists claim long-age for everything from Creation to
evolution, where typical outspoken six-twenty-four-hour-day
Creationists claim short-age for everything from Creation to the
flood. But both are partially wrong. Are the cataclysmic flooding

events that we see evident world round somewhere in between the two debate positions, oh say, hundreds of thousands of years rather than millions? No, they were not.

Reading the Facts

From an evolutionist perspective, and I suppose from an atheist perspective as well, evolution is based upon a wild and unprovable assumption of very long-ages, and those long-ages require morphing or evolving us and the Creatures into our current forms. Without long-ages, the ever-expanding big bang would not be nearly as widespread as we see the heavens today; but what about the Creationist perspective?

Are those six-thousand-year time constraints realistic? For the flood, yes, a flood occurring about forty-five hundred years ago fits perfectly with the available evidence as discussed throughout this book, yet flood supporters almost always misread the facts of both geology and of the Bible. The flood obviously occurred just as it is stated in the Bible and as detailed in this book, there is no other logical realistic explanation for the widespread similarities in geological strata. And for anyone who would dare to invent any theory that one continent would get flooded without the others being flooded also is quite honestly utterly irrational, and such beliefs are nonsensical. Similarly, the observable data that is available to us in the heavens screams of a slow Creation. If you truly believe in God and that this "God" created all things, then you are not going to foolishly assert that the Earth is only six-thousand years old by cheating God's own glorious Creations and God's physics just to imagine that your six-thousand-year-old Earth belief is correct.

Christians need to understand that they do not have to accept the big bang if they accept long-age Creation, neither do they have to accept evolution or big bang if the Earth formed billions of years ago. A very old Earth has no bearing on the flood or the deposited layers of flood sediment that we now see.

Let's try to put this all into perspective. The Earth is very old, likely older than even the big bang supporters claim it to be. But here is a critical juncture that can clear things up for Creationists with regard to Creation *and* the flood. If you have read *The Science Of God Volume 1 - The First Four Days*, you should be aware that the "days" of Genesis One are quite obviously ***not*** twenty-four-hour days, and if you read *The Science Of God Volume 2 - Day Three - Gravity, Land, Seas, and Evolution of Plants* and *The Science Of God Volume 3 - Day Five and Day Six - The Creatures - Revolution or Evolution* you will understand why.

The most important part to grasp here is that any *Creation* and *flood* debate points are, in truth, in no way connected to each other. Nor is the evidence that is found between the sediment layers connected to Creation. It is only because evolution supporters insist that those layers "prove" an evolutionary progression of life that the subject of layers deceitfully enters into the arguments surrounding Biblical Creation. And that is perhaps the biggest hoax ever perpetrated on the world. This is why if the layers are ever undeniably proven to have been rapidly laid down by a worldwide flood, then evolution theology is utterly destroyed forever! Evolution ***requires*** those layers to be long-ages between them, and also each layer to be very old. The entire evolution model is built upon that long-age deception. And that belief-set is also connected to big bang long-age creation, which all combined together creates a great amount of confusion for those who hear the information but do not take the time to carefully and accurately study it for themselves.

But let's use the evolutionary approach and examine the layers and the type of life found within the layers. It is claimed that the lower in the stack of layers that a layer is found, then the more primitive the creature will be that was trapped by that layer. For instance, most of the lower layers have simple marine life in them, the types that crawl around on the bottom of the ocean. Then as we climb the layer-ladder and look through successive layers, the creatures become more evolved—more complex, until we get near the top where we finally find modern man.

But is there another possible explanation for all of this? Could there be an explanation that would fit the layers with the flood *and* explain the apparent evolutionary progression of creature-forms as the layers were deposited one upon another in rapid succession—without cheating science? Yes, there is an explanation, and it is far more plausible and far more logical than the theory that long-age layers are "proof" of an evolutionary process.

You are now at the point in this book where you need to pause and take a breath, clear your mind, and be able to decisively separate issues in your mind—The Flood and Creation have no affiliation with each other whatsoever. Yet, many people mistakenly and inadvertently blend the two events. Much of this mental blending of events has been caused due to the evolutionary agenda that has been pushed onto the world over the past century or two. Pop-science's Darwinian end-to-end evolution is a creation topic and it *requires* the sedimentary layers that were actually laid down due to the flood to be long-age layers. These layers are found all around the globe. Without the long ages, evolution could not have occurred as proposed by evolution enthusiasts. You might want to read this paragraph over a few times to settle things in your mind regarding the confused connection between the flood and Creation, confusion that was caused by heavy indoctrination of evolution theory that was taught as "fact". **Grasp it, it is very important to understand!** This mental association occurs because evolutionists take the layers produced during the flood events and strongly assert that over very long ages the creatures progressively evolved, which is of course based upon how evolutionists interpret the layers.

When trying to wrap your mind around the evolution-versus-flood aspect of the entire debate, you must first and foremost realize that the layers were laid down very rapidly in comparison with evolution's long-age, millions and even billions of years perspective. You also need to grasp the fact that evolution is a challenge to both Biblical Creation **and** to the Biblical flood. If the evolution side is correct, then the Genesis One Creation account is wrong and the Genesis Six and Seven flood account is also wrong, because then the animals are shown to have evolved throughout the successive layers, rather than having being created on Creation days five and six as laid out in Genesis One. And if evolution is true, then it demands the layers to be much older than the flood suggests, thus invalidating the Creation, the flood, the entire Bible, and God along with them.

Let's consider the reality of what would happen if the scenario of a supporting rock matrix collapsed beneath the continental plates. First, we have to acknowledge that this would cause a great deal of rock, stone, and silt to be forced to the surface with the force of the extreme pressures that the water is ejected with. In addition to that, we would expect massive seismic and volcanic activity to quickly work its way through to the surface as friction heats up the base of the solid body of landmass at random locations sitting atop the rock-matrix near any fractures in the continental plates that were caused from the shift that resulted from the crumbling matrix. All of this extreme geological activity would, without question, cause many massive tsunami waves. These waves will travel across the oceans at varying speeds depending upon the severity of the particular seismic event.

We would also expect that due to any neighboring supporting rocky matrix collapsing, the next area of matrix in sequence will give way to the increased continental load that it was then bearing. This would cause a succession of geological activity that could go on for many years, decades, or even centuries. Then when massive tsunami waves sweep across the land, we have even more weight from the water and silt deposited on the continents causing more failure of the supporting matrix structure. In the case of a crumbling supporting structure we would expect the tsunami wave to carry sediment across the land

and deposit it in relatively even layers of varying thicknesses. These layers would include varying mixtures of sedimentary stone, silt, and volcanic material with some layers being more dominant with certain types of material, and other times having a general mix of the materials. The materials caught up in the tsunami waves were quickly carried across any land that the tsunamis traversed and then the materials caught up in the tsunami were deposited into layers via the material dropping out of suspension.

Everything in what was just stated would logically occur upon failure of the supporting rocky structure beneath the continents. Now for the interesting part; evolution asserts long ages between layers, but all erosion patterns and all fossil evidence indicates that the layers were laid down in relative rapid succession. If you look at a place like the Grand Canyon, you will see that it is a massive washout area. While rushing water is powerful and often very violent, no amount of water is going to cause the Grand Canyon to wash out in the manner it is if most of the layers are made of solid rock. However, if the layers have not yet cured to a fully hardened state, and if they are in a putty or soft rock state for a long enough period of time, then it wouldn't take much to wash the canyon out. The canyon is best explained through actual logical reasoning, rather than guesses of long ages.

It is the assumption of long ages that has become the veil that shrouds our understanding about places like the Grand Canyon. The question with the canyon is, if the canyon was carved out because of the flood, then when and how did that occur? It is unlikely that the Grand Canyon was formed during the Biblical flood year. However, it was formed because of subsequent flood effects. Many of the layers were likely deposited during the forty days of the flood and even for many weeks and months and even years after.

We will get back to the Grand Canyon shortly, but first let's explore the fossils. When the layers of sediment were being deposited, which creatures do you imagine were least likely survive the recurring tsunamis and deposition of sediment? Would it have been larger animals, or small slow sea creatures that typically creep along slowly on the ocean floor? The obvious answer is that the smaller water insects and shellfish would have been the first to be trapped in the sediment. The large animals and birds would have been able to swim or fly to keep their heads above water much longer than those other creatures. As soon as the tsunami waves went past the larger creatures, many of them would have once again found ground to rest on. A tsunami wave would pass by in a matter of minutes or even seconds. As the weakest creatures began to tire from the successive tsunamis, they too would eventually succumb to the sediment. This same pattern is found all the way up the layers. However, there are anomalies in this that the evolution supporters will not acknowledge; this includes anomalies such as further developed life being found in some lower layers.

The strongest smartest, most resilient creatures would have survived the longest, thus enabling them to endure long enough to get themselves trapped in some of the upper layers. We also would expect some large sea creatures to have been carried along in the tsunamis, and if those creatures either succumbed to the silt in the water or simply got trapped on the land as the tsunami receded, then they would be out of place making it appear as if inland lakes once existed for a very long period where their now dead carcasses lay. Then, with some of the large creatures now lying dead on the surface, subsequent tsunamis would bury them in yet additional layers of silt. All of this perfectly matches the actual physical evidence that we see all around the globe, and it is all highly plausible and has been witnessed in modern tsunamis. We also must realize that there would have been regional tsunamis and worldwide tsunamis. With as much seismic activity as would have occurred for the flood events, there would have been near constant tsunami waves occurring, with them ranging from very small to very massive with all of them carrying sediment and creatures along with. Some creatures will have been trapped *within* a layer rather than being trapped *between* layers.

Now we have a vast array of creatures sifted throughout the layers and organized by each creature-type's survival capabilities as the layers progressed. The flood waters are now calming down somewhat, and our Earth is a ball completely covered in water and floating in space. But even though the worst is over, the seismic activity is still occurring. As the continents start to drift, portions of them begin to rise up, eventually forcing the land up through the water. Noah and family can now exit the Ark with feet firmly on the ground. Some raised areas of the world would have more layers and some less, and some layers would remain mud for years and others would allow the water to rapidly seep out enabling that ground to very quickly "dry".

If you ever wondered how Earth's aquifers came to be, the secret is in the flood. The waves would have carried stone, sand, and rocks along with them and deposited that material along the way, but since that type material will settle out almost immediately, those larger and less sticky items would have gotten pushed along and clustered into any low-lying areas that eventually were covered by additional layers of silt, sand, and other dirt material. This is similar to how a typical beach works as it changes from day to day as the water splashes upon the shore, but in the case of flooding or tsunamis, when the material is moved along into a low-lying area, it gets trapped there.

Let's also take note that during the first several weeks after the forty days of flooding, that many of the drowned creatures' carcasses not yet buried by sediment would have been bloated and would have been floating in a vast ocean, this would have been occurring all around the world with many animal carcasses. When the bloating dissipated, then the carcasses would eventually descend to the land beneath the water. The floating creatures all eventually sunk to the bottom on top of roughly the same final sediment layers.

Let's back up a bit and consider the many footprints found around the world. There are many areas with foot prints of creatures that are pressed into rock layers. Our logic and common sense tells us that at one point the material had to have been soft enough for footprints to be impressed into it. Now, if these layers took millions, thousands, or even a year or two in order to be deposited, then what are the chances that the foot prints would have eroded completely or at least substantially before any subsequent layers were deposited? Those chances are about zero. We can pretend that a cataclysmic event occurred that suddenly covered those footprints, but then evolution is confirming rapid deposition of sediment. No material that must be in a putty-state in order for footprints to occur is going to have the foot impressions in it survive for more than a few months, if even that long, without them being destroyed or substantially damaged. Also, there are allegedly areas of exposed rock that have footprints from both dinosaur and man in close proximity. If such reports are true, then it would prove that man and dinosaur co-existed.

What became of the creatures that had bloated and stayed afloat until the bloating dissipated? Those creatures would have shown greater decay as they lay exposed lying on the land but still immersed in water. These creatures would likely have had some of their flesh eaten by carnivorous sea creatures that had access to them, but their bodies also would have been more susceptible to decay, especially after the water receded from those areas after the tsunami wave had dispersed.

Still Too Much Water

All of the things just mentioned in the last section are consistent with what we would expect in a flood largely caused by continental collapse. But we are still stuck with one overwhelming problem, which is that of the massive excess of water left after the flooding stopped increasing. Sure, we can legitimately lift the land a bit in one area to explain the Ark being able to sit on dry land shortly after a global flood. But there is no possible way that all of the water is ever going to evaporate into the sky or soak into the soil in order to expose all of the land that we see today. And making up stories about asteroids striking and boiling the water off into space are childish, as is any notion of that much water seeping back into the Earth in any way. This presents us with some questions that we must answer. First, if the land rose up, what caused that to occur? Next is if the land was raised by continental movement, then did the Ark sit atop something like our modern-day mountains? And finally, exactly how tall was the tallest mountain?—what was its pre-flood altitude?

Let's take a look at the last question first. There is a notable difference between the **altitude** versus the **height** of a mountain. Altitude is based upon sea level, which changes due to tidal pull occurring due to the monthly alignment of the Moon and Sun—altitude is a global measurement. Height, on the other hand, is a local measurement. If the majority of the land mass is 300 feet above sea level and a mountain on it is 800 feet high, then the mountain is 800 feet high with an altitude of 1,100 feet. But if the same mountain is placed on a land mass that is 1,000 feet high, then the top of that 800-foot mountain now has an altitude of 1,800 feet. Water finds its gravity-level based upon altitude rather than on local measurements. The Ark could easily have been on a high hill or mountain and theoretically have had an altitude lower than the plains area of another continent. What we don't know is the basic average land level just before the flood began.

For now, let's look at land movement. The questions regarding land rising up under the Ark need to be answered. We know that modern-era witnesses, and even Darwin himself, have given testimony that land can quickly rise or fall due to seismic activity. But if the continents moved and the land buckled, then wouldn't the Ark have ended up on top of a mountain? No, there was certainly movement, but if a three-mile-thick piece of continent was horizontally compressed only fifteen cubits (25 feet), then it would not raise the entire continent evenly. Realistically, it would compress the softest or weakest area and raise it about fifteen cubits.

To try to picture this: take a 25-foot-wide slice out of North America from top to bottom, from northern United States to the Gulf of Mexico, and cut it all the way through the thickness of the continent. And then push the continent back together to fill in the cut-out area. Now lay down the 25-foot-wide slice flat on its side on top of the land right where the slice was cut out, and now you have a 3-mile-wide by 2,500-mile-long strip of land raised 25 feet above the rest of the land. Gently compressing land of a continent could easily and quickly raise the land far more than the required 25 feet or fifteen cubits. The increase in land elevation will be proportionate to the size of the compression area together with the horizontal distance compressed plus how evenly the plate is compressed. At the point in time where the Ark was just about to sit on dry land, the atmosphere would at that time generally already have absorbed all the water it could. Since the Ark may have been resting on solid ground, but was still in the water, due to little more than, oh let's say, fifty feet of sideways continental compression, then with only a bit more of a continental push, the Ark is now on many miles of dry ground. Such a small amount of movement could have put them hundreds of feet above water within days, giving them miles and miles of area to roam free on and begin anew.

While all of that is a very a plausible and logical explanation of Noah and family's entry into the new world, we still have that one pesky nagging problem of—where did all the rest of the water go to? It is at this point that we need to do a bit of speculating about the pre-flood land altitudes. This takes a bit of thinking to follow because you must keep several thoughts in mind all at the same time. First, we don't know the level/height of the pre-flood land masses. The entirety of the land could have been relatively low and flat with only minor deviations of a couple hundred feet. This would mean that the underlying rocky matrix might have only crumbled down a hundred feet or less depending upon topography. As the land descended, the water level would have risen at the same time, as was mentioned in an earlier chapter.

Regarding increased water height, since the supporting rocky matrix was not a complete void, but rather only a percentage of void filled with water, the water expelled from the "great deep" would not have had as great of an effect on water levels as the land descending into the water would have had. If the void beneath the continents was approximately fifty percent total void area, then based upon the current land and water surface areas of the entire globe, for every foot of drop in land altitude the water could have risen one to two inches due to water being ejected from beneath the continents.

This is not sounding good for the plausibility of the Bible's flood. However, we still have to calculate the ocean rise as the water was displaced by the land that had previously protruded *above* pre-flood sea level. Since the current oceans cover roughly 70 percent of the face of the Earth, that means that for every two feet that the land descended into the water, the water rose just under a foot. So, subtracting the 50 percent rock from beneath the plates we would end up with something close to a 1 to 3 ratio. The land goes down three feet and the water raises 1 foot giving the appearance that the water had risen a total of about 4 feet. Thus, if the supporting rock matrix collapsed three hundred feet worth of height, then the water would have come up on to the land four hundred feet from its pre-flood water level relative to the land.

It's easy to see how easily the flood occurred, especially if the old-world land was anything like the Plains areas are today, which are very broad and very flat. We also have to note that land with hills is thicker than land without hills, so those areas might have descended deeper bringing the high spots closer to the level of the surrounding land. But, there are simply too many variables to be able to make solid estimates of land elevations both before and during the flood. Everyone must realize that the old world looked nothing like the topography that we see today anywhere around the world. While this explains in very basic and logical terms how the flood worked, it still fails to explain the disappearance of the water after the flood waters stopped increasing.

In our efforts to explain the disappearance of vast amounts of water, we cannot simply accept such illogical assumptions that somehow the land is magically floating or that a catastrophic event occurred boiling away the water, or that all of the water was absorbed into the Earth.

This means that we have a real water problem regarding the worldwide Biblical flood! It just doesn't seem logical at this point, but let's continue anyway.

Chapter 23

The Right Questions

Our biggest hurdle in trying to find any answers is asking the right questions. For instance, asking how the flood waters covered the mountains of our time is not the right question. A better approach is to ask, how and when did the mountains form? The first question wonders how the water rose to the heights of our current mountains, and thus immediately traps us into having to explain a whole lot more water than was ever realistically available. But the second question is closer to the point, as it is wondering how the mountains got there, and as a second aspect to the second question, also *when* they got there. This should ultimately lead you to the proper question, which is, were our current-day mountains even there at all before the flood?

So, as we work to try to get to the final resting place of the massive amounts of water that covered the globe and its unexplainable disappearance long after Noah and family left the Ark, we have to figure out what the proper question is. Once we find that question, we will either reject the Biblical flood or we will reject full-scope evolution. The questions that we must ask as we seek our answers are either going to be *how* questions, or they will be *why* questions. All too often Christians get far too hung up on the *why* aspects when they are expected to deliver *how* answers, so they instead offer more of a *why* answer of "God did it, God can do anything!" This might be true, but it does not offer any useful answers to our *how* questions, and in doing so it ignores offering the scientific explanations that people are seeking and quite honestly deserve as to how God did things and how things occurred scientifically.

Our questions arise from the things we wonder about, so when you try to reason through the answers for evolution or Biblical flood questions, keep your focus on the *how* aspects. You can credit everything to God if you would like, but then the question becomes, *how* did God do it? When we give God credit for magically doing everything, we then tend to lose focus on God's physical facts, so it is best to set God's interaction aside. God often initiated things and then let things play out in a physical manner, which is what happened during the flood. God initiated the flooding and really had no work to do afterwards because God's physical laws established during the Creation events discussed in the previous volumes of *The Science of God* would dictate all activity thereafter.

Our goal is to ask any questions we can think of in order to find the current hiding place of the many millions of cubic miles of water that went missing sometime after Noah and family left the Ark. And we are going attempt to do this without invoking any additional interaction from God. If we cannot find out what happened to the water, then the flood is a farce—plain and simple!

Climate

There are many people who offer opinions about their view regarding climate conditions before, during, and immediately after the flood. With our understanding of climate from our modern-era Earth, we know that large bodies of water hold heat and moderate temperature, thus reducing temperature fluctuation.

After Noah and family left the Ark, nearly the entire globe was covered with water. We know from basic logical scientific analysis that more than twenty feet of water could not possibly have been absorbed into the air via evaporation due to the saturation limits of the atmosphere. So, we must assume that the land lifted beneath the Ark in order for any dry ground to have appeared that quickly. There simply is no other logical scientific explanation. Maybe not all, but most of the rest of the land would have remained under water at that moment in time when they were finally able to exit the Ark. With a nearly completely global ocean, the waters would have kept the air temperatures quite comfortable everywhere on Earth, even on the cooler poles.

The heat generated from the crumbling continental understructure and from the pressures and movement of the ejected water would have warmed the oceans considerably during the flooding, possibly killing off some marine life, but also making a perfect sea climate for other marine life. Many sea creatures would have thrived in perfectly warm temperatures and likely would not have had many predators. We can speculate that in the oceans you will find a sudden die-off of marine life beneath the ocean floor. And along with that, there was also sure be dead bloated land creatures that were carried out to sea that sank to the bottom of the ocean and were subsequently covered by layers of sediment.

The waters of the ocean likely developed temporary currents due to tsunamis and continental activity keeping the water in motion for many months or even many years. The climate on Earth at the time would have been quite stable and topical. Noah and family lived on an island in the early days after the flood. This environment offered the Ark's animals the perfect environment that they needed to prolifically thrive.

No Waves

Many flood theorists and evolutionists insist that the Ark would have to have been built super strong because they insist that the weather would have been severe hurricane-type weather with very rough seas. This implies that the entire crew experienced forty days of extreme weather, and then occasional storms in the nearly eleven months following the forty days of flooding. But, since we don't know those weather conditions, no one can say for sure what the conditions were. However, considering that the Ark was allegedly on an area with a higher elevation, they might not have experienced any rough seas at all, at least until they were lifted off of the ground by the flood waters.

Being on a hill or mountain at a somewhat higher elevation would have kept them free of the initial massive tsunamis. Additionally, if they were located more centrally on the "super continent", which they likely would have been, then the tsunamis would have lost most of their power before ever reaching the Ark. As the water level increased, the water being ejected from beneath the continents would have been losing some amount of power, causing a reduction of the severity of seismic activity and tsunamis as time progressed. The Ark may have experienced a fairly calm voyage the entire time. But with the reliable stability of the boat design discussed earlier in this book, the only thing that would be a danger to them would be to be broadsided by a massive cresting tsunami wave.

Now, the Ark in its empty state would be extremely stable in any amount of turbulent seas. But a fully loaded ark is another story. In one aspect the Ark would be somewhat more stable regarding small waves when fully loaded. But the dangers of a fully loaded Ark could at minimum tip the ark onto its side if weight shifted. Normally the Ark would right itself, but as every shipmate or cargo hauler understands, shifted cargo can have devastating effects on the stability of the vessel carrying the load.

In rough seas, a loose load can be thrown to one side inside of a vessel that is caught in turbulent conditions. When that occurs, then everything inside will become even more tightly packed to the side it is listing towards, causing the problem to become worse until the vessel completely capsizes. Depending upon the construction of a vessel, a shifted load can capsize the boat, but if it is designed properly then the shifting will be limited.

"Make thee an ark of timber planks: thou shalt make little rooms in the ark, and thou shalt pitch it within and without." The important part of this quote from Genesis Six is "thou shalt make little rooms in the ark" These "little rooms" are the key to the Ark's stability. If the animals were in several rows of little rooms all along the length of the Ark and were adequately penned in, they would not have been able to shift completely across the Ark. Being in their "little rooms", the animals might have been tossed against the walls of those "little rooms" in very rough seas, but other than that, no major shifting of weight could have occurred because of their being confined to one side of the Ark or the other. In addition, logistically, the larger animals would have been loaded on the lower decks along the outside walls, thus completely balancing and stabilizing the load. The Ark was unsinkable, and if it did capsize momentarily from a massive wave, it would likely immediately have righted itself if the cargo space is mostly sealed so that water cannot too quickly enter the Ark.

Rough seas often don't have massive cresting waves out at sea; instead they are typically massive swelling waves when away from land and shallow waters. Since the Ark would have been adrift, they would likely not have been cutting directly into the waves. This would make a big difference in how the boat would handle rough seas. And since the world was entirely covered with water for a while, there would have been little opportunity for creating waves because severe weather is greatly increase by landforms, and cresting waves are often caused by the waves' energy hitting the land underneath shallow waters.

No Weather on the Boat

Since the waters likely calmed and slowed as the forty-day deluge was coming to an end, it's not stretching anything to assume that they might have experienced very little turbulent waters, or the Ark may have been on ground until the last ten to fifteen feet of the flood lifted had the Ark off of the ground or if the land dropped beneath them. When the flooding finally ceased, the Sun would likely have shone shortly thereafter, revealing a sky rained clear, with tropical temperatures all around the globe. In fact, with the temperatures moderate around the globe, they probably didn't really experience much strong weather afterwards at all, although they would still likely endure some tsunami activity. But if you're on the water in your Ark away from land, a tsunami is not really much of a problem because out in the ocean, tsunamis are somewhat gentle rolling swells that would simply raise the Ark while the tsunami energy passes beneath it.

While they might have encountered severe seas, severe seas are not a prerequisite at any time during any part of their voyage through the entire Genesis flood. Their entire voyage could easily have been fairly gentle and very stable. We can however, expect that they would have felt and heard the Earth quaking beneath them. While the Ark was strong enough to endure rough seas, it probably didn't really endure much, if any, rough seas at all.

Chapter 24

Ask and You Shall Receive–The Missing Water

When God speaks to someone, they can consider it a promise or a covenant. You can be sure that whether you follow or defy God and continue living life following or defying God's promises, you will receive what you have earned; this eternal Truth will not short you in your deserved pay. This includes misleading people by teaching false and inaccurate thoughts about the Bible as if those thoughts are true. This is true even if you are innocently doing so in purity and in all sincerity. If you accidentally feed someone poison, you are still partially responsible for their death. And if someone tells you that you are feeding someone poison and you continue to do so, then you are a murderer and fully bear your own guilt for your own deliberate evil behavior. How much more if you cause the demise of someone who abandons God due to faulty teaching? Always be very careful in fully understanding what you are teaching others.

Lack of understanding is the biggest problem that most people encounter. All too often, we hear what we want to hear and we then latch on to those thoughts while never even bothering to challenge them. Or we choose to defy things that we hear, but never really try to investigate to see if those things might actually be true. Sure, we might read up on something, but then we immediately make our prejudicial decisions about *what we think* and we never really get to the root of the issue. It all comes down to our lack of will to try to understand any given information. This is true in relationships *and* in physical circumstances like the Biblical flood.

You cannot fully understand anything unless you ask. In the New Testament Jesus The Christ said, "Ask, and it shall be given you: seek, and you shall find: knock, and it shall be opened to you. For every one that asks, receives: and he that seeks, finds: and to him that knocks, it shall be opened." Christ was telling us to open our hearts, minds, ears, and eyes so that we can understand. In Hosea it says "My people have been silent, because they had no knowledge: because thou hast rejected knowledge..." Many Bibles say "lack of understanding" instead of "no knowledge" Knowledge and understanding are connected but they are not the same. There are many people with great knowledge, but they have no understanding. Knowing something is good, but understanding it allows us to excel with that knowledge.

If you work to understand something, you will then seek more than the basic facts, you will seek to know the purpose and function of those facts. So in regard to *evolution, Creation,* and the *flood,* most people opining in these thought-areas all examine **the same exact data**, examine **the same exact layers**, and read **the same exact Bible**, and yet the differences in their conclusions could not be further apart. The frightening part about this is that, typically, both sides are so far off the mark that they are both able to dismantle the other side's debate points, and yet it does not dawn on either side to re-evaluate their own theories after their theories have been brutally dismantled by the debating opposition.

We must not only ask, we must also *understand* so that our follow-up questions gravitate toward finding the Truth of the events that we seek to understand. Always seek to understand

before drawing your conclusions. Understanding begins with asking the right questions, and asking the right questions begins with understanding. This seemingly impossible problem is easily overcome by simply dropping our prejudices while approaching the topic with a free mind having only the quest for Truth at heart. It is then that we should ask our questions and be willing to hear the actual true answers that might be different than what we want to hear.

These Numbers

It is important to note here that while some people might debate the specific numbers used in the calculations throughout this book, the basic concepts are non-negotiable and are simple observational fact. You can change up the numbers all you want, but all of the principles are simple physical math that any student or math teacher can play around with and achieve varying degrees of success with. Also, some points are certain, such as all of the water is still there, and the principles stated in this book regarding the water did occur at some point in Earth's past. Whether it was only several thousand years ago or millions of years ago is of no matter regarding these numbers. While the numbers might need some adjustment for decimal accuracy and land and water mass estimates, what you read here occurred as is made quite evident by studying Earth's land and ocean floor topography, regardless of when it all occurred.

So, Where is the Missing Water?

The ultimate question, regarding the evolution versus Creation and the flood, is about the water. Did water cover the entire Earth? Or is the Bible wrong? If it did cover the entire Earth, then where did the water go after the flood?

We can lay down theories of how the underlying rocky matrix structure below the continental plates collapsed and that the rocky matrix structure below the continental plates is the "great deep" mentioned in the Bible. And we can imagine that when the Earth formed, that there was molten rock near the surface which was porous much like lava rock, and we can also consider vast networks of caves that lava once flowed through and that most of those structures collapsed, thus spewing forth the now pulverized rocky material in the form of boulders, gravel, and all sizes of grains of these materials in the form of sand and silt, all having been made as such on their way out of the great deep. This all seems plausibly logical.

We can also assume that the land was very flat and had a low altitude relative to sea level before the flood. And we can theorize that all ice melted and that all rain storehouses in the sky were emptied. We can also assume that from the combination of those things the entire globe was covered with no less than fifteen cubits, or about twenty-five feet, of water. We can even go so far as to assume that seismic activity caused the land to lift beneath the Ark and that in a matter of days Noah and family had scores of square miles to run aground upon on which to re-establish life. But with all of those key points that are likely to have been actual factors in the Biblical flood and the subsequent relatively fast appearance of a bit of dry land, you still need to clearly understand and remember that the vast amount of remaining water was still there flooding most of the Earth at that point in time. And we're not talking about only a few feet of water here. No, we are talking about hundreds of feet of water above most of the land.

To try to solve the mystery of the missing water, we first have to ignore the altitude of our modern mountains. The mountains matter, but regarding the missing water, we are not trying to find the amount of water that would have filled the Earth to 15 cubits over that height. Rather, we need only to calculate using the equivalent of the **average** land height. This is the area that is so problematic for most people who ponder this topic. In fact, even some of the most brilliant scientists miss this point because they assume that the continents moved over millions of years, and therefore they imagine that if the Bible's flood account is true, then, in their rationale, our modern-era mountains were covered by at least fifteen cubits of water, which is an absurd thought. But even if we remove the mountains we still have over a thousand feet of general land height to deal with. And yet we have dry land today, but all of the flood water is still here unless God somehow used hocus-pocus magic to remove it. So, where did it all go?

The Mystery is Solved with Volume

Similar to calculations you might attempt regarding the Ark and using different cubits and/or different wood types etc. to find the available payload volume and weight, the numbers in the water mystery can also vary substantially depending upon the amounts of land and water you start your calculations with. However, it again is important to note that the basic premise being put forth regarding the land and water and all things involved with those calculations will not fail accurate math. There are different ways to approach the math, but only one way is going to offer an accurate perspective.

Today the average land elevation above sea level, including the mountains, is about 2,778 feet. Keep in mind that the 2,778 feet is not some mathematical averaging of various height measurements from around the world, but rather it is taking the land-volume above sea level and distributing it evenly on the current exposed land area of about 57.5 million square miles.

Multiplying the total land area on Earth of 57.5 million square miles times the 2,778 feet or 0.53 miles gives a total land volume above sea level of 30 million cubic miles. The total volume of the Earth's oceans is about 315 million cubic miles having an average depth of about 2.26 miles. With a nearly 11:1 ratio of ocean volume to land-volume-**above**-sea-level, the oceans could easily accommodate all the exposed land volume many times over.

Earth is about 196.6 million total square miles, and of the 196.6 million square miles of the globe, water covers just over 139 million square miles and land covers 57.5 million square miles. But we don't really care much about the **square** miles of land at this moment in our calculation; we are mostly interested in the **cubic** miles of land at this point in our calculations. Again, the average land height above sea level is 0.53 miles which is the 2,778 feet above sea level as just mentioned. To arrive at the cubic miles of land, we have to multiply the half mile of average land elevation which is 0.53 miles, times 57.5 million square miles of land surface area on Earth, giving us a total of about 30.2 million cubic miles of land above sea level in Earth's current topography of today.

The ratio of square-miles-of-water to square-miles-of-land is about 70-parts-water to 30-parts-land (2.3 : 1). The amount of land currently protruding from the water would displace nearly an equal amount of water. (The land would absorb a negligible amount of the water, but it is not worth considering in our calculations).

With the surface area ratio of the ocean to land being at 2.3:1 or the 70:30 water to land ratio, if we could cut off the land level equal to the water level and then put that cut-off part into the ocean, then the extra 30.2 million cubic miles of land would raise the water level of the ocean about 811 feet. We arrive at the 811-foot increase by dividing the 30.2 million cubic miles of land by 196.6 million square miles of Earth's total surface giving 0.1536 miles of added ocean depth. That is to say that we multiply 0.1536 by 5,280 feet in a mile giving us the 811-foot increase in ocean depth. This means that if the average land volume above sea level was placed in the ocean, doing so would increase the ocean depth about 811 feet.

Now, since we theoretically cut the 30.2 million cubic miles of land above sea level off of the land mass, the land's surface level is now equal in height to sea level. But since the ocean level and land level are now the same, the amount that sea level increased when we put the cut-off dirt into the ocean is also increased over the land area as well. This means that the newly trimmed land surface is now underneath 811 feet of water.

At this point in our calculations it is important to note that once the land surface is at the same level as the water surface, then nothing we do will increase sea level. So, when we placed the cut-off portion of the land into the ocean it increased the water level 811 feet all around the globe, including over the land. So now we need to go back and reduce the amount of land to cut off and find the point where the amount of land being cut off and placed into the ocean will bring sea level up equal to the level to which we cut the land down to. This is referred to as the point of equilibrium. With the numbers that we have been using in our calculations, the amount of land to trim off of the top is 1,966.51 feet. If we could cut the upper 1,966.51 feet off of a uniformly even land mass and then place that into the ocean as an extension of the existing continents, then the water level, that is to say, sea level, and the surface of the land would be equal. At this point the water will not increase in height at all, even if we could push all of the land far below the surface, this is because all of the land is already displacing the maximum amount of water.

But, God obviously didn't chop off the land and cast it into the sea, so we first have to realize that ***nothing magical*** occurred and **all of the water and land is still here on Earth with us today** otherwise it is all a great big lie. However, regardless of what it all proves, all of the evidence is sitting around us out in the open for all to see. If we divert from that understanding at all, then we may as well give up because we can then invent any nonsense our minds can contrive. Such diversion is the sort of rationale that brought about both the six-twenty-four-hour-day Creation theory and the long-age evolution theory. This must stop!

If God is real and did create everything, then we cannot cheat true science, logic, facts, or evidence. Evolution supporters pretend that they are "science", but that is obviously wrong. And the six-twenty-four-hour-day-Creation supporters have an opposite tendency to attribute everything thing to a magical-hocus-pocus God who ignores his own inventions and cheats them to make things appear as if they are something that they are not just to satisfy their own lack of understanding.

The evidence is clear in both the Heavens and on Earth, and the only thing barring us from seeing the truths are the biases that we each harbor and refuse to let go of.

Now that we have made sea level and land surface level equal nothing can be added or removed. And, again, we can force the land as far under water as we want, but sea level will not change at this point. So, since we have taken the dirt we removed from the land and used it to extend the continents, we now must calculate how much larger the continental surface area is. To do this we take the amount of dirt that we trimmed off of the continents which was 1,966.51 feet or 0.372 miles and multiply that times today's land surface-area of 57.5 million square miles giving us 21.4 million cubic miles of dirt removed. Now, the ocean depth or thickness of the continents is 2.26 miles, so we now divide 21.4 million cubic miles by 2.26 miles giving us an added continental area of 9.47 million square miles. Our aim here is to calculate the total approximate pre-flood continental surface area. At this point it is simple; it will be a figure less than our added 9.47 million square miles of continental area added together with today's continental surface area of 57.5 million square miles. We take 57.5 million square miles plus 9.47 million addition square miles, which equals about 67 million square miles of pre-flood continental surface area.

We simply do not know and likely will never know the exact square miles of pre-flood land. So then, what's the significance of the added 9.47 million square miles of land surface, or even the calculated pre-flood land surface area of about 67 million square miles? And what has it all to do with the 57.5 million square miles of land that we live on today?

Before we can answer that, you must realize that **all** of the land is now flooded to any amount you want to force the land beneath the water's surface, and sea level will not rise. If the pre-flood continents were 67 million square miles then where did that missing land go to? Did it sink into the ocean?

More Mountains of Evidence

Even if we take the extended land area of 9.47 million square miles and force it deep into the ocean, the rest of the land would still be flooded or equal to sea level. Now, since long-age science believes the mountains are millions of years old, and that false long-age information has been taught for a very long time, it causes many people to be deceived when they see remnants of the past in coal mines. Many coal mines are buried deep in the earth, even to a thousand feet deep, which makes sense in regard to a global flood, but how do we explain the coal found at the **top** of some of our current mountains? Coal is allegedly a collection of organic matter that was highly compressed for extended periods of time under high heat conditions, yet near the tops of mountains it would not have been compressed enough to form coal—unless... the mountains were once not mountains.

Coal veins are an important source of energy that holds the mysteries of the past secure. Many people fight the use of coal as an energy source, yet it is an abundant and inexpensive and clean energy source when used properly. As a side note, ask yourself: Why are they so afraid of using coal?

Now, let's take all of the points discussed and put them together to see if we can answer that pesky question about the apparent disappearance of 21.4 million cubic miles of water that has been displaced by the 9.47 million square mile extension of the continents.

Regarding the flood, we need to first realize that the face of the Earth from before the flood would be unrecognizable to us today. The land back then was likely considerably flatter to a point of being more plains-like everywhere. As the porous supporting rock began to give way, the continents collapsed raising the water and lowering the continents at the same time. This allowed for rapid flooding and explains the "fountains of the great deep", and the seismic activity explains the layers that were caused by tsunami waves rapidly depositing silt, and while so doing, trapping nearly all of the creatures in silt at some point during the flooding period and for a time afterwards as previously discussed.

Due to the tumultuous geological activity, about a year after the flood began the land shifted and caused some of the land to be rapidly forced up allowing the Ark to sit upon dry land once again in a very short amount of time. The land rose far enough to expose plenty of land for Noah and family and all of the animal passengers of the Ark to exit and once again dwell on the land. Some evaporation occurred, but that is minimal in the bigger picture, so we still need to somehow make the vast amounts of water vanish without invoking any hocus-pocus-God-magically-did-it nonsense.

The water mystery is completely unrelated to the flood. This is because once the flood was completed and the land flooded the flood is irrelevant to our solving the mystery of where the water went. So, in these calculations we are starting with a completely flooded Earth.

We have to focus on two critical points: The first is to assume that the flood actually did occur just as the Bible states, and that **all** of that same water still exists today here on Earth. Thus, we are faced with arriving at a logical explanation of the apparent of discrepancy of land versus water. We found that if we submerged todays exposed land that it would raise the ocean approximately 811 feet above current sea level, yet we are not able to explain where the water went, although you have been given hints.

This discrepancy is easily explained through cubic miles, versus square miles. The volumetric displacement of the cut-off portion of the land is 21.4 million cubic miles and the current surface area of the land is about 57.5 million square miles. Since the average ocean depth is about 2.26 miles, we divide the 21.4 million cubic miles by 2.26 to arrive at 9.47 million square miles of surface area that is 2.26 miles thick/deep equaling our 21.4 million cubic miles of land volume.

Since the land is currently approximately 57.5 million square miles we have to add 9.47 million additional square miles to the 57.5 million, which brings us to a possible total of about 67 million square miles of pre-flood land surface-area before the continents collapsed. So where did the missing 9.47 million square miles of land vanish to? We already established that once all of the land is submerged, the water will not rise any more no matter how deep we mathematically force the land beneath the surface. And since we already forced all of the land under or equal to the water's surface, we have to find a legitimate way to reduce the water level and increase the land height.

This is the most important part for people to grasp in order to be able to easily picture this and readily visualize the solution. Since the land was approximately 67 million square miles before the flood in this mathematical exercise, we have to find the percentage of land that is now apparently missing. The missing 9.47 million square miles divided by 67 million pre-flood square miles is a 14.15 percent difference. This means that 14.15 percent of the pre-flood land volume is nowhere to be seen... or it is?

The answer to the missing land is hidden in the mountains and highlands. For this, you have to have access to a topographic map showing the mountains. When looking at the maps you have to examine the topography, most of which is extremely obvious, and then decide in your own mind if you think that those obvious landforms reveal land shrinkage of nearly one seventh of all land area. The answer is clear when you study the mountainous areas.

The following example is over-simplified, but should give you somewhat of an idea of what actually occurred. In the years following the flood there was a great deal of continental movement. The ocean floors show clear signs that something moved. When you follow the residual markings on the ocean floor, you will see that they show that the Americas and the African continent were at one point attached together. Many flood supporters will attribute that movement to the forty days of flooding, which is possible, however, the continental drift more than likely transpired through more than a hundred years, and occurred through at least until sometime after Phaleg's time when "the earth was divided". All of the land would not necessarily become exposed at the same time. As the continents moved into their current positions and heights, that movement would have changed the overall weight balance of the Earth which would explain some of the differences we believe are detected in Earth's gyration or axial wobble.

As each continent moved toward its current location, the entire mass of the continent would have been moving in the same general direction. As the continents moved, the leading edge of the movement was gaining friction against Earth's foundational crust. And then as this happened, the mountains began to form, resulting in increased downward pressure causing even greater friction between the continental plates and Earth's foundational crust. This slowed the movement and increased the mountains in width, length, and height. As the continents slowed down, the compression forces from the friction on the leading side of the continent together with the momentum from the trailing end, created astounding amounts of pressure all along the land causing a large portion of the surface area to raise slightly. These physical realities are the same whether you use long-ages or Biblical flood-ages during calculations.

Key to the Missing Water

Here is the key to solving the mystery of the missing water/land discrepancy: As the land and mountains began to rise, the volume of land that was exiting the water was and is directly proportional to the square miles of compressed surface area multiplied by the continental height below sea level. This level of geological activity undoubtedly caused even more tsunami activity. However, the early tsunamis would have flowed over the new growing mountain ranges, such as the Rocky Mountains while they were beginning to form. The tsunamis that happened well after the Ark was firmly on ground would also have deposited sediment from possibly below the continent, but also from the sediment that had previously covered the mountains before they began to form.

As the tsunami waves washed over the very low but growing mountain ranges, many of the previous layers of sediment were now lifted up on top of the growing mountain ranges and still were quite soft. As the tsunami water washed over them, those additional layers were re-dispersed into the tsunami waves and subsequently deposited in the non-mountainous plateau areas. This explains the inconsistent nature of most of the fossils found today. The dead bloated creatures that floated immediately after the flood would eventually completely de-gas and fall to the submerged surface of the land. These carcasses would have sat in water and then eventually on dry land as the waters receded due to lateral land-compression forces raising the land surface.

When the later tsunamis flowed over the mountains, the soft layers that covered the land that was now slowly becoming the mountains were washed away and mixed with the water to once again become silt that would settle as yet more layers covering the once bloated creatures. As the continents moved toward their current position, the seismic activity from continental drifting was reduced. When the mountains first began to form and were still fairly low in height, the water from tsunami waves that made it over the forming mountains was trapped. This created massive areas of inland water, some of which are where we now see desert areas. So, we would expect to find sea creatures as well as dinosaurs and other creatures in those latter formed layers.

Eventually, through seismic activity or simply water flow, one of these giant inland seas made its way to the oceans and took with it what once was a massive plain of newly deposited silt and sand, and in so doing left behind what we now call the "Grand Canyon". If those layers that make up the walls of the Grand Canyon were as solid back then as they are today then the Grand Canyon would not be what it is today, and there would likely still be an inland sea upstream of that area.

If you travel across North America to the Appalachian mountain range you will notice that the area is considerably smaller and the mountainous average height is considerably lower than their western rocky counterparts. This is consistent with the compression we would expect from the momentum of trailing mass of land east of the mountains including the part called the continental shelf.

The ideas revealed throughout this book need to be divided into two separate categories. One category is making unprovable speculations such as Noah and family could have had a garden on the entire roof of the Ark during the year immediately after the flood-waters stopped rising. The other category is the actual evidence and the logic that matches that evidence.

When it comes to explaining the disappearance of the flood waters, we really are only concerned with the 21.4 million cubic miles of land that was horizontally compressed by about 14 percent or $1/7^{th}$ the total pre-flood continental surface area. If we assume that the entire Earth was flooded at any point in the past, whether forty-five hundred years ago or forty-five million years ago, the same problem exists. And since the entire globe shows evidence that it was underwater at some point, a global ocean is nearly absolute. Time is irrelevant as to the question of the missing water. So as stated earlier, all of the water still sits on the face of the Earth. It was the land that changed due to compression from continental drift. And as the land compressed by about 14.15 percent of total pre-flood square miles in this mathematical exercise, the water dropped and the land rose resulting in today's topography. This is true if it took one-hundred years or one-hundred-million years to move—the math principles do not change one bit.

Try It Yourself!

Take a large container, preferably made of clear glass so that you can see what is happening. An aquarium works great for this experiment. Now, add about 3 inches of damp well-packed sand to one end of the aquarium and make sure that it is very smooth and level on top. At this point slowly add water to the aquarium so as to not disturb the sand and mess it up. Keep adding water until the sand is saturated and the water is level with the sand with no sand sticking out of the water at all but very close, about two paper thicknesses over the land. Once your water is saturated into the sand and your water level is at the level of the top of the sand with no sand out of the water, take a board that is the width of your aquarium container and push it to compress the sand sideways about 1/7 the area of the sand and see what happens!

It is important to place a mark on the glass exactly at the water line before you push the sand so that you can see the water level drop and the sand surface level rise when you do this experiment.

If you have a 12-inch-wide aquarium that is 30 inches long, then you would have an area of well-packed sand that is 3 inches thick and 12 inches wide by 10 inches sitting on one end of the aquarium. And then after the water is in and settled, as just described, and you have placed a mark on your aquarium at the water level, then push your board to compress your sand the full 12-inch width and compress the entire 12-inch width about 1 1/2 inches, making your 10 inches now only about 8 1/2 inches.

When doing this, it is very important to be careful to not have too much water in the aquarium. If your water is well above your sand then the experiment will not be accurate to what happened after the flood. Again, to scale this thought close to flood values using 3-inch thick sand, the water should be no more than 2 paper thicknesses above the sand.

After your initial observations, you can compress the sand even more to get a clearer idea of exactly how this works as you watch the land rise while the water level drops while compressing your sandy land-mass more and more.

In the following illustration, the triangles have been placed on top of the land. Removing the triangles lowers the water in direct correlation with the volume of material that was removed. If we took all of the highlands from all around the Earth and cast them into the sea right now, then the entire globe would once again be flooded. It is the compression amount of the land that both lowers the water and also raises the land. In the case of the flood of Noah's time, the current highlands and mountains of today are equal to the Biblical flood water volume that covered the land. This same situation is true regardless of when this occurred whether forty-five-hundred years ago or forty-five-million years ago; the math principle does not change.

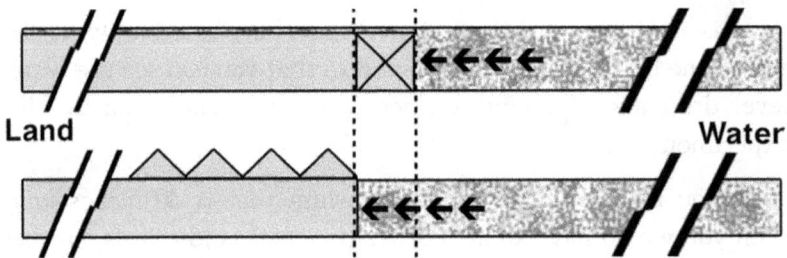

Figure 12. Land compression forced up mountains and lowered oceans.

Math Perspectives

It seems certain that there will be those who will doubt these numbers and try to make this out to be nonsense, but again whether forty-five-hundred or forty-five-million years ago, *when* this occurred is irrelevant. The math exercise you just read is solid and it works. In the exercise, we took 1,966.51 feet of today's 2,778 average feet of land above sea level and cast it into the ocean and it raised sea level 811 feet making the land surface and sea level equal in height.

We then very successfully compressed the land to pop up a few mountains, and then the problem was solved! This ferociously proves out the basic mathematical concept and is an undeniable mathematical truth. What you have to ask yourself is, it is possible, that based upon the current topological evidence from all around the globe, that the land that we see today was horizontally compressed as much as 14.15 percent giving us today's continental topography? Look at a topological map and ask yourself, "Could the mountains and highlands be stretched out about 1/6th the total continental area?" If so, then you have the answer.

The table of numbers below shows the formulas and numbers used for the calculations late in this book. The formulas shown at the end of each line will produce the figures on each line. The bold italicized lines are the only lines where you can change figures. All other lines will calculate based upon those bold italic lines and the formulas on each line that have formulas included.

196,900,000.00	**A.) *Earth's total surface area***
139,382,879.00	**B.) *Earth's current ocean surface area***
57,517,121.00	C.) Earth's current land surface area (A-B)
5,280.00	D.) Feet in a mile
2.26	**E.) *Today's average ocean depth (Miles)***
315,005,306.54	F.) Earth's ocean volume in cubic miles-NEVER changes (BxE)
2,778.00	**G.) *Today's average feet of land above sea level***
0.53	H.) Today's average miles of land above sea level (G / D)
30,261,848.89	I.) Today's cubic miles of land above sea level (C x H)
1,966.51	**J.) *Feet of today's exposed surface land to put in ocean***
811.49	K.) Remaining feet of land above sea level (G-J)
21,421,968.49	L.) Cubic miles of today's exposed surface land being put in ocean ((J/G)xI)
8,839,880.40	M.) Remaining cubic miles of land above sea level (I-L)
811.49	N.) Feet Box J or L increases sea level based on Box B 140million (L/B)xD)
574.44	O.) Feet Box J or L increases sea level based on Box A 197million (L/A)xD)
0.00	P.) When ZERO, Box K equals Box N, then land and sea levels are equal (K-N)
9,478,747.12	Q.) Added land surface area due to Box J or L (L/E)
66,995,868.12	R.) Estimated PRE-flood land surface area (C+Q)
14.15%	S.) Percent of PRE-flood horizontal compression of land to achieve Box C (Q/R)
1/7	T.) Fraction of PRE-flood horizontal compression of land to achieve Box C (1/S)
16.48%	U.) Percent of POST-flood horizontal stretch of land to achieve Box R (Q/C)
1/6	V.) Fraction of POST-flood horizontal stretch of land to achieve Box R (1/U))

Figure 13. Table of figures used to calculate land and sea levels.

When considering this information, you must remember that in reality, all of this would have been considerably more nuanced because the pre-flood continents were resting upon the rocky matrix below the continents, thus raising the pre-flood continents up out of the water. When the rocky matrix collapsed, all of the land was submerged. Then as the continents began to drift/shift, they started to horizontally compress due to the horizontal movement. Then the land began to become exposed as it was forced up out of the water due to that horizontal compression, which simultaneously began to lower sea level equally by volume due to the volume of land exiting the water.

Does the Bible Pertain to Everyone?

In the last section, if you are honest with the data, it is very apparent that all of the land on the globe was at one point covered with water due to flooding, or flooded due to the continents dropping. It is also apparent that the continents moved. And it is mathematically accurate that removing approximately 21.4 million cubic miles of the 30.2 million cubic miles of the above sea-level land would make up the space of 9.4 million square miles of additional land if the 9.4 million square miles of land was 2.26 miles thick, which is the depth of the ocean. And if we submerged the 9.4 million square miles of land down into the ocean, sea level would rise about 811 feet, thus making the water and land surfaces equal.

Those are all basic mathematical facts. Whether the land was flooded all at the same time cannot be absolutely known, but with the overwhelming evidence of the similarity in layers and where creatures are generally found within those layers, it logically indicates a worldwide simultaneous cataclysmic flood. However, the time and duration of that flood are not known. Flood supporters say forty-five-hundred years and evolution supporters who are flood deniers indicate any such catastrophic activity anywhere near that level would have had to have been tens of millions of years ago if it did occur.

Major catastrophic events are undeniable all around the world. What we need to do is to try to see if we can determine the time period in which such event(s) may have occurred. To try to get to the bottom of this, let us take another look at the mountains that hold so many secrets. In our modern era, we mine coal from various sources. It's all from the Earth, and if we are correct, it is the result of large collections of organic matter being swept into large groups and then buried for long periods. The question is how long of a period elapsed before it became what we now call "coal"?

Here is a little-known secret about coal: Coal is preserved energy that was heated to fairly high temperatures but not allowed to ignite, thus it could not burn. In other words, trees and grass and animals were for some reason all pushed into confined areas and then buried and heated for some period of time. The little-known secret is that you can make real coal in weeks or even days. If wood is placed inside of a very strong sealed container and heated to temperatures around the ignition temperature, then it will not and cannot burn because it has no oxygen. And if the container allows pressure to escape then the moisture can escape as the temperature passes the boiling point of water. As the temperature of the enclosed wood increases, it will blacken, and with the water removed and heat and pressure applied, the wood will compress. When the wood is cooled to the ambient temperatures and the container is opened, you will find a piece of coal inside. This process is a very short process. It does not take decades or centuries or any length of time more than the few necessary days, however, the coal that we mine today likely developed over decades.

When organic matter is confined tightly and it still contains moisture, then as the material begins to decay it will heat up to the ignition point. You can witness this yourself if you visit a fall leaves compost area when the residents all bring their leaves and pile them up. After a week or two you will see steam and smoke coming from the piles of leaves if the problem is not dealt with—

this is coal in the making. In fact, in the twentieth century when many farms still built wooden barns where they stored the animal feed and bedding hay in the upper part of the barn, if they failed to allow the hay to dry properly before bringing it into the barn, then occasionally it would heat up and spontaneously combust causing a barn fire, something fairly common during that era.

As you can see by the basic evidence, trees being tightly grouped together and then buried and compressed would begin the fermentation or decay process which produced a great deal of heat. Since the organic material is buried, it does not have access to oxygen and therefore cannot ignite and burn away. The heat will continue until the material exhausts the bioactivity and turns the water to steam, which is then expelled and forced through the sediment that lay on top of the large clustered mass of trees and other organic materials. Since these areas of coal are so large, the coalification process was not going to occur in only a few weeks, but rather likely over many decades due to the enormity of the material clusters.

The relevance of coal is that some coal is under ground and other coal is found at higher altitudes inside of mountain tops. These coal veins were deposited and coalified around the same time. Logically, if the flood occurred as described in the Bible, then trees and vegetation would have been swept away by the tsunamis and possibly even from shockwaves from fracturing continents. But tsunamis are the most likely cause of the clusters of trees, animals, and other vegetation. If you have ever had the opportunity to watch a flash-flood come down a hilly region, you will have a very good idea of how this works. As the tsunami wave comes onto the land, it picks up more and more debris. Then, as the tsunami continues to sweep across the land, the debris at the leading edge of the tsunami increases its ability to clear the land of *anything* in its way. If the trees at that time were large trees that were anywhere near the size of the great redwoods on the west coast of the United States, then a massive tsunami would have plowed those trees down like a bulldozer through a raspberry patch.

Much of the organic matter was pushed into massive clusters during the forty days of flooding which likely would have occurred early during that time, and then as the sediment came along, the newly gathered clusters of organic matter were buried. When the Appalachian Mountains formed, the coalification of those clusters of organic matter had already been completed. And when the continent was completing its movement, the hills were formed with the coal veins already inside. Vast amounts of time are not needed for the deposition of layers or for the making of coal.

To take this further, people who work on excavation, around low-lying wet and swampy areas where the soil is very fertile, are well aware of a somewhat oily look and smell that is often experienced. Sedimentary organic matter collects far faster than we want to scientifically admit, and when it is clustered for long periods, it can also become what we call "crude oil". If you think back to the continental movement discussed in this book and think about the organic matter at the bottom of the oceans, you should be able to piece together where deep-well crude oil might have originated from.

Does any of this relate to the Bible beyond the flood? And does it matter to you? Well, that all depends on your personal choices. If you don't believe in God then I suppose not. But if you choose to ignore the abundant evidence that lay before your very eyes every day that you live and breathe, and if you ignore every point made in this book regarding full-scope evolution, versus Creation and the global Biblical flood accounts so that you can believe that there is no God, then you might have a problem.

We can choose to believe that what we believe is true, but we cannot change what is *actually* true. The flood happened and this book walked you through every critical point, each of which can stand alone without depending upon some obscure theoretical point that many people doubt. Evolution and long-age coalification and long-age tectonic movement all ignore facts and evidence and are assumptions based upon modern-day indoctrination about misinterpreted observations. What science does today regarding tectonic movement is like looking at a wildly erratic stock chart and taking a snapshot of a single brief moment at the end of that chart and then extrapolating it back in time on the chart and assuming that the entire chart followed that graphic trajectory. This is not how anything works in the free-flow of nature or in the free-flow of real life.

The Bible is a very accurate book when read without the poison of our human predispositions. It is scientific and teaches scores of lessons about the ramifications of the unseemly behavior of man. If we don't want to repeat the errors that occurred before the flood, then we all had better wake up and work to better understand the actual words of the Bible in order to avoid our own demise. If we choose to believe the Bible is not true, then we are denying Truth, and that is not a forgivable sin. So, are we all doomed? Will God flood the Earth again to wipe out our filth and cruelty?

The Rainbow at the End of the Pot of Gold

We often hear the phrase about "the pot of gold at the end of the rainbow", but it appears to be the other way around. The "pot of gold" is the information and understanding that is wrapped up in a world of layers of physical data for you to examine in order to know the Truth. The value in that little pot of gold... well... really Earth-sized pot of gold, is your eternal existence. Will there be more *Volumes of The Science of God*? Possibly, but none will be more important than are the first five volumes. What we all must decide for ourselves is if we believe in God. If we decide that there is no God, it will have no bearing on the Truth. So, if we are right, then we are right, but if we are wrong then there will be hell to pay. What we all need to ask is: What is a soul? And, is it eternal? According to the Bible it is eternal. A "soul" is not scientifically detectable, yet it exists. There are people all around the world with various body parts missing from accidents or from birth anomalies, and they still live and often thrive—they are *who* they are. A soldier that served and lost his legs is still the same awesome man he was before he lost his legs. His soul is not in his legs, and the same is true of any other body part, including parts of the brain.

Our bodies are shortcuts to the process of soul-making or soul-awareness and are nothing more than that. A soul is made at the moment of conception and is taught full awareness through living life. Killing a person, whether in the womb or later in life, robs that soul of some of its awareness. Can a soul become more aware after death? Maybe, but according to the Bible we can be reasonably sure that the soul will either dwell eternally in regret or eternally in joy. If God *does* exist and heaven and hell are real, then we have to ask ourselves if we believe that souls are eternal. If they are eternal as the Bible appears to indicate, then living in the torment of regret because you cannot enter Heaven is a very heavy eternal price to pay. But lucky for us, God made a promise to never again suddenly destroy all living creatures.

Even though the people of Noah's time were warned, they ignored that warning and continued in their evil ways, some even mocking Noah and family for their strong faith and Ark-building venture. But as the story goes, that did not work out so well for those people. As a matter of fact, it is very possible that some of their bodies are our fuel today as we mine coal and drill for oil. As promised by God, we now can be assured that we will not pay for our neighbor's sins when *everyone* drowns in a global flood, instead we are all individually culpable for our errors and we will pay dearly for those errors if we do not change our ways.

The environment of the Garden of Eden that Adam and Eve were removed from is not the same as the pre-flood environment. In as little as about sixteen-hundred years after Adam and Eve were expelled from the Garden of Eden, humanity had corrupted itself so terribly that it was deserving of certain death. Even though vehemently denied by many people, evidence of giants does exist which further proves the Biblical account of that part of history.

The evidence of a global flood is everywhere and the physics math proves that it could have occurred exactly as stated in the Bible. The only valid explanation for the water's ability to flood the entire globe is the fragile supporting-structure in the "great deep" being "broken up" under the continents. Anywhere you dig deep enough on Earth, you will find signs of the Biblical flood.

The Ark construction methods are not specifically known, but the design offered in this book can easily be tested by you through building a small-scale model of the Ark. There would have been plenty of food for the crew and for any animals that did not go into some level of hibernation. By the end of the voyage there may have been thousands of extra creatures on board that were borne from any creatures that have short regeneration and gestation cycles. Food was likely stored in abundance for Noah's family and the animals, but they also could have been growing food on the roof of the Ark if needed, or went fishing in their free-flow pool-basin area in the base of the Ark.

The lateral compression of 9.23 million square miles of the continents allowed for the rapid exposing of our current landmass without having to attempt to explain the water away through evaporation. You can easily calculate the values when looking at global maps to see were the land was obviously horizontally compressed. The Ark was raised up out of the water due to horizontal compression of the continents, allowing Noah and family to relatively quickly and safely set foot on solid ground once again!

God made a vow to man after the flood. This was not just any vow, it was a rain<u>vow</u> that God would never again destroy all living things with a flood. In Hebrew, the letter "B" or "bet" can also be pronounced as "vet". Interestingly the rain-**vow** that God made to man was in the form of a rainbow seen after it rains and that vow sits atop the vast pile of evidence that we call Earth.

Choose the Rain Vow
After
You examine the Earth-sized Pot of Golden Evidence!
Thank You God, for Your Rainvow!

UNDERSTANDING THE CHURCH

Upon This Rock I Will Build My Church

Church in the Lurch - a House Built Upon Sand

The Church is rapidly dying, and much of the clergy in recent times have been doing it more harm than good. People are fleeing from the Churches as they seek a religious perspective that fits a modern worldview. Should we revive this old Church and try to save it from its own demise? What exactly is "The Church", and who or which of the many religions is the official caretaker of it?

The Christian religions of the world have done their fair share of damage to themselves and to the world, but in the bigger picture, they have done more good than damage. Saving the Church is probably worth our collective efforts because the Churches are perhaps the most charitable group of organizations that existed throughout history and even up to today.

The main reason that the Churches are in the rough condition that they are today is due to a lack of understanding by clergy and congregation. We can overcome this dark era of the Church and revive it only through *Understanding The Church*.

Understanding The Church will help you in Bible study, or even to simply better understand the Church. But most importantly, *Understanding The Church – Upon This Rock I Will Build My Church* will help to revive this dying patient.

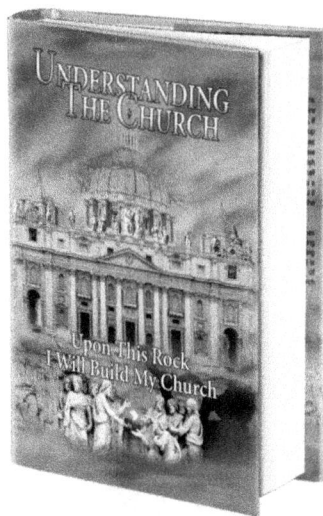

Search: Understanding The Church Book

Volume 1 - The First Four Days

Is there a God? Did we evolve? Did everything start from a big bang? These questions have been plaguing our minds for many years. Only science-minded people and clergy seem to have the answers. But do they really have any true answers?

Is what we are told by science true? Is what we are told by the Church true? Or are there other better explanations for everything? Did we hitch a ride from Mars, or is that all fantasy science? Was everything created in six twenty-four hour days, or did it all take billions of years to happen? Few people are willing to even fully consider these questions, and even fewer have any coherent answers. *The Science of God Volume 1 – The First Four Days* challenges your current beliefs while asking tough questions of science and of the Church.

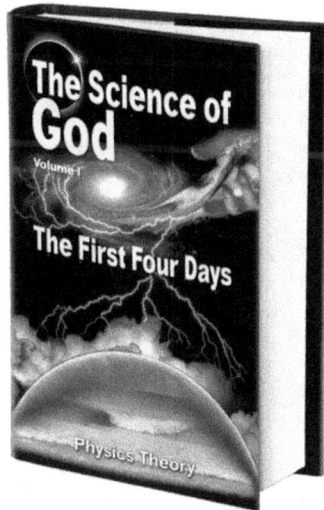

For years, Christian after Christian has attempted to argue for God and the Bible's Creation only to fail miserably. Why is this, why is it that Christians cannot seem to win this debate? Often Christians think they are winning the debate only to find themselves at a loss to answer the real questions, and then they get mocked for their poor answers.

Whether you are a scientist or an average Christian and want to discuss the Creation debate, *The Science of God Volume 1 – The First Four Days* is a mandatory read for you. *The Science of God* takes you through the thought process to enable you to speak intelligibly about Creation, the cosmos, evolution, and astrophysics.

Search: The Science Of God Book

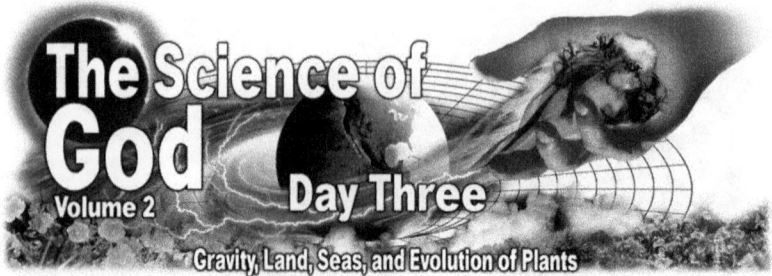

Rocking the Cradle of Life
A Decent Account of Descent

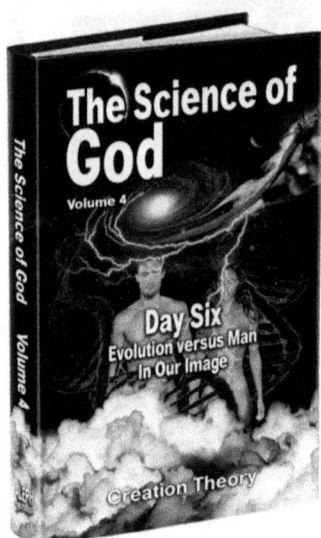

Have you ever wondered if humans actually did evolve from apes? Or maybe, if we were specifically created, then how might have that occurred? There sure are a lot of opinions on the evolution versus Creation topic. And too often these views use confusing technical jargon that few people care to learn or have ever even heard.

The answers to the questions you might have are, in many cases, the same answers that many other people seek. When you have solid answers that are difficult for someone to thwart, it's good to share those answers so that others can also feel confident with their own understanding of the arrival of mankind and the level of importance that it has in their own lives.

The Science Of God Volume 4 - Evolution versus Man — In Our Image takes a deep but simple dive into the human evolution versus human Creation debate using simple language that everyone can understand and enjoy!

If you have thoughts that you have been reluctant to share, then suspend your thoughts for a bit and open your mind to consider the perspectives and evidence presented in *The Science Of God Volume 4 - Evolution versus Man — In Our Image*. You will acquire a much clearer view of the subject as you read the various points made in this engaging book about the arrival of mankind.

Search: The Science Of God Book Volume 4

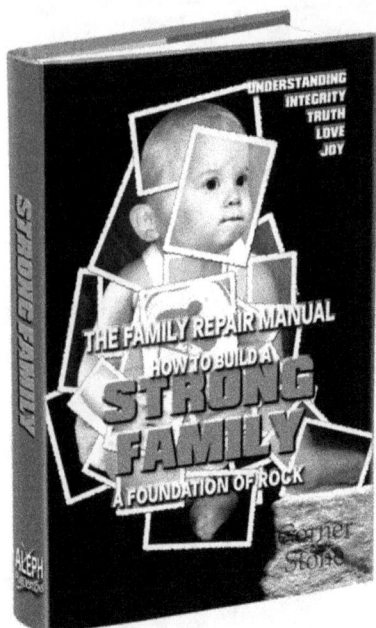

MARRIAGE MANUAL
MAKE YOURS A
Red Hot Marriage
Made In Heaven Filled With Passion and Joy

Learn the Secrets to a Successful Marriage

Have you been trying unsuccessfully for years to tell your spouse the way you truly feel? Are you suffering in a lackluster marriage? Is your marriage on the rocks? Are you planning on getting married in the future? If you answered yes to any of these questions then *Red Hot Marriage* is for you! This straight-forward book covers these and many other common marriage problems and also reveals the causes and solutions for some problems that are not-so-common.

The information in this powerful book, like a true friend, can be at your side with each step you take in restoring your life and relationship to where you likely imagined them to be.

We all deserve lives filled with joy and passion, but our relationships have been tainted by society and by our upbringing. *Red Hot Marriage* strips away all of the lies that we have been inadvertently taught, and quickly teaches you how to regain control of your marriage so that it can be as robust, fulfilling, and passionate as you expected. The mysteries unveiled in *Red Hot Marriage* can have you in command of your marriage in short order as friends and family watch in amazement while you and your spouse walk the path to a strong, vibrant, healthy *Red Hot Marriage*!

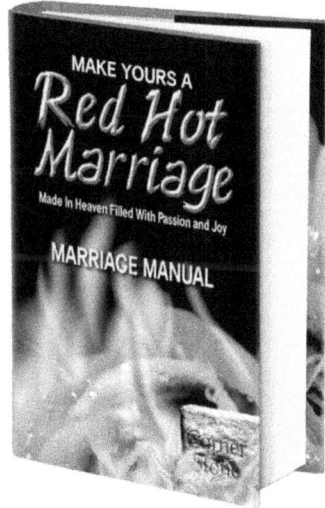

Search: Red Hot Marriage Book

Notes

Notes

www.ingramcontent.com/pod-product-compliance
Lightning Source LLC
Chambersburg PA
CBHW071413090426
42737CB00011B/1449